Understanding Cultural Landscapes

Understanding Cultural Landscapes provides an introduction to the range of approaches characterising the study of landscape and culture. This textbook discusses the origins, underpinnings, and practical applications of social science and humanities scholarship concerning landscape and its theorisation.

The book offers an easy-to-digest introduction to relevant approaches and ideas in the study of landscape. The text includes practical, real-world examples drawn from a range of social science and humanities disciplines to showcase the interpretation of landscape by different means and to different ends. "Pauses for thought" encourage the reader to reflect and connect ideas and debates to examples encountered in their own life and work. Suggested activities support independent study and aid the design and delivery of university teaching. Each chapter also includes suggested readings, and the textbook includes discussion of research methods for those developing their own research projects.

This book will be valuable for both students and educators encountering, or teaching, the study of landscape within the humanities and social sciences for the first time. The incorporation of practical exercises and suggested reading enables the book to supplement independent study or be used as a teaching aid.

Rachel Hunt is a Senior Lecturer in Geography in the School of GeoSciences at the University of Edinburgh, Scotland, UK.

Ben Garlick was a Senior Lecturer in Geography at York St John University from January 2017 to October 2024, during which time he taught a module on 'Culture and Landscape'.

"In times of increasing socio-environmental conflicts and crises, landscape has become a crucial but also contested terrain for exploring our interplay with the material world. *Understanding Cultural Landscapes* provides a unique and insightful guide into the ever more complex field of cultural landscape studies, spanning from traditional approaches within human geography to its current critics and the most recent theoretical explorations in the field. With its clear structure and pedagogic layout, it offers a much needed introduction into the field."

Professor Mattias Qviström, *Professor at the Department of Urban and Rural Development; Landscape architecture Spatial planning, SLU the Swedish University of Agricultural Sciences*

"This excellent, wide-ranging and accessible text is a joyous celebration of landscape's 'unruliness'. It guides the reader through landscapes for analysis, landscapes in process, landscapes felt, experienced and lived. Drawing on an impressive variety of geographers, historians, philosophers, poets, literary critics, feminist thinkers, decolonial theorists and more, this book will help any student think differently *about* and *with* landscape."

Dr Pauline Couper, *Associate Professor in Geography, Associate Head: Geography and Liberal Arts, York St John University*

"*Understanding Cultural Landscapes* is a significant statement of renewal. It considers why a key concept continues to matter in human geography and cognate fields. Its co-authors have compiled an open and searching treatment of landscape, demonstrating a readiness to unsettle as much as secure what is known. The insights and challenges they offer make this a critical text for students and educators alike."

Professor Hayden Lorimer, *Chair of Human Geography, University of Edinburgh*

Understanding Cultural Landscapes

Rachel Hunt and Ben Garlick

Routledge
Taylor & Francis Group
LONDON AND NEW YORK

Designed cover image: Getty images

First published 2026
by Routledge
4 Park Square, Milton Park, Abingdon, Oxon OX14 4RN

and by Routledge
605 Third Avenue, New York, NY 10158

Routledge is an imprint of the Taylor & Francis Group, an informa business

British Library Cataloguing-in-Publication Data
A catalogue record for this book is available from the British Library

ISBN: 9781032000411 (hbk)
ISBN: 9781003172444 (pbk)
ISBN: 9781003172451 (ebk)

DOI: 10.4324/9781003172451

Typeset in Times New Roman
by Newgen Publishing UK

For our partners Lou and Doug, and our wee ones Briar and Oran, Emmy and Fraser.

Contents

Figures

Boxes

Acknowledgements

Thanks to our students for learning landscape with us. Thanks to those many teachers who stoked our enthusiasm in this concept. And, finally, thanks to those who read versions of, commented upon, and immeasurably improved the content which follows.

Introduction

Understanding Cultural Landscapes

The Landscapes of our Teaching

When this textbook was first imagined Ben lived and worked in York. While there, a favourite local place to visit was the Yorkshire Sculpture Park (YSP). From 2022, after we had written of its potential for introducing approaches to the study of cultural landscapes, he began taking students there too (see Garlick and Hunt, 2022).

Located near Wakefield, in northern England, UK, YSP offered Ben and his students a real, material "landscape" in which to think about what that term meant in the context of cultural theory: i.e. as a *cultural landscape*. It is, as the Cambridge Dictionary (2024) defines "landscape", certainly "a large area of countryside, usually one without many buildings or other things that are not natural; a view or picture of the countryside". Yet such a definition hardly encompasses all that YSP is, the ways in which humans have shaped its environment, or what it *means*. Established in 1977 within the grounds of an eighteenth-century country estate (Bretton Hall), the park includes nearly 100 artworks set amidst 500 acres of grassy meadows, woodlands, and lakeside (About YSP, n.d.). Spending days there with family and friends, and latterly students, it was almost as though one could walk, on foot, through the pages of the textbook we were writing. Here, in this place, the many interwoven, complementary, and contradictory theorisations, applications and 'cultures' of landscape could be played with, turned over, and better grasped.

Thinking about how to begin this textbook, we were struck by the number of cultural landscapes we have visited over the course of our careers teaching in higher education. These have been places to teach *with:* providing a lens to talk about, complicate, or translate key approaches in cultural landscape studies (as well as other areas besides). The YSP, Saltaire industrial village and World Heritage Site, Dalkeith Country Park, Jupiter ArtLand, York's inner-city St Nick's nature reserve, the beaches of Nairn, the raised mire of Humberhead Peatlands near Doncaster, the island of North Uist, and urban centres in Belfast (Northern Ireland) and Ljubljana (Slovenia) have, amongst others, been variously called upon to ask (and answer) the questions of interest to social science and humanities scholars of landscape: how, and by whom, were these landscapes *made*? How are they maintained or transformed? What do they mean? What do they *do*? These are all questions central to *Understanding Cultural Landscapes.*

DOI: 10.4324/9781003172451-1

Such questions variously concern relationships, vision, perception, power, sensation. They also relate to themes of identity, politics, economic, and knowledge. Sometimes we are brought to things that are extraordinary or unusual; at others we notice the components of tacit, everyday experience. Encountered via what Donald Meinig (1979) called "[t]he beholding eye" of the landscape researcher, each place becomes a kaleidoscope as field-working and teaching quickly explode a single, often taken for granted or seemingly settled landscape into a complex cultural entity. As we have written regards the possibilities of the YSP for understanding this concept, "[l]andscape names a construction of our cultural imaginations; the accumulated traces of overlapping labours; a domain of bodily experience; a field of material forces; and a space charged with feeling, memory, and emotion" (Garlick and Hunt, 2022: 247).

This textbook is testament to the recognition that "landscape" names no simple or singular thing (see Box 0.1). In introducing *Understanding Cultural Landscapes*, we hope to set its kaleidoscope in motion. Inevitably, as we discuss below, this is a textbook born of a specific time (what we understand of the foci and cultural landscape scholarship *now*), as it reflects our own arrival into the debates we consider (we are both geographers, though interdisciplinary in outlook). The continuation of interest in our own field and others in landscape as a vital, relevant and *useful* term to think, research and write with is what underpins this project, and we hope that it helps you (the reader) recognise its value and potential applications in fathoming the relationships connecting environments and cultures.

Box 0.1 Defining "Cultural Landscape"

To a significant extent, our use of the term "cultural landscape" reflects the origins of landscape study itself. Thus, in Chapter 1 we discuss how a "cultural landscape" was originally conceived and deemed an important object of empirical interest for social science and humanities scholars (principally geographers). It is important to note that we use this term from a desire to cast the scope of this book's readership more widely, as many of the topics discussed herein are of interest to the diverse field of "landscape studies" in its social, political, economic, and human dimensions.

Throughout this textbook we avoid over-defining the meaning of "cultural landscape", not least because a central aim is to emphasise that there is no single meaning in use. Instead, we concentrate on what this term serves to *connect* in terms of ideas and approaches. Nevertheless, since clarity around the use of terms is important, we therefore present two ways in which the term "cultural landscape" is frequently used or discussed within the book.

1) "cultural landscape" understood as a material entity or spatial formation. A cultural landscape constitutes an empirical (real world, observable) object as the outcome of human interactions in a specific environmental and spatial context, especially where such interactions have transformed (either physically, in terms of meaning, experientially) that environment or our relationships with it.

2) "cultural landscape" understood as a conceptual category or designation. Here "cultural landscape" may be the term used to indicate that an environment, setting, or place corresponds to certain criteria or characteristics that are held to be associated with this term and is therefore to be valued or framed as such for the purposes of management, governance, or organisation. It may also, however, be a phrase applied to signal how an area or space is being interpreted, defined, or argued about in the context of a given theoretical or conceptual approach.

Between these two uses – as thing and idea – we demonstrate how, across the sheer diversity and variety of what can be called a "cultural landscape", there exist multiple approaches that frame "landscape" in relation to specific aspects of social and cultural theory in order to focus on and attend to particular aspects of human–environmental relations. This wide-reaching, undisciplined, and undefined nature of landscape is held by many to be a strength, and we encourage you to be always alert to how culture, landscape, and cultural landscape are being understood and applied across the chapters that follow.

Chapter Outline

As an introduction to the textbook, this chapter addresses a range of anxieties that exist around studies of cultural landscape: their relevance, partiality, and unruliness. We do so to make the case for cultural landscape as a focus: why it matters and the contribution it can bring as a focus and a concept in researching human–environment relations. We say something of our own positionality to be transparent about this book's basis, as well as potential areas of omission. In closing, the chapter sets out the content of the book to come.

Why Landscape? Anxieties and Promises

We are mindful that, for some, "landscape" is an old-fashioned term. For others, it is overly limited or limiting. And for still others, it has been castigated as vague and ill-defined to the point of being useless to the scholar of human–environmental relations. In this brief section of the textbook's introduction, we want to make the case for landscape in relation to three areas of potential *anxiety* that the term can induce. These are, firstly, a sense of the irrelevance of landscape in a time of unlocatable, multi-scalar environmental crises and cultural phenomena; secondly, the sense that a term which finds its roots situated in Euro-American scholarship and assumptions may not be appropriate for articulating the diversity of cultural–environmental interactions now, or indeed in the past; and thirdly, that the unruliness of landscape makes it unwieldy and unhelpful in articulating a project of scholarly interest. We address each in turn, both to acknowledge the potential limitations of a cultural landscape approach, but also to signal its abundant promise.

Relevance: Landscape in a Time of Crisis

There is the sense that "crisis talk", and concern with crises, have long been central to humanities scholarship and its self-definition (see Reitter and Wellmon, 2021). It is also evident how often we are reminded that we live amidst some or other new crisis requiring of urgent understanding to be remedied. Over two decades ago, feminist philosopher Chris Cuomo (2003: 97–98) in a keynote address asking us to consider what useful, ethical knowledge for a "better world" might look like in a time of crisis, observed that we are all, by virtue of the relations that define our existence, "members of economic and environmental communities too large, too diverse to even imagine". In this context, the agency of individuals acting upon their environments "is less like a vector and more like smoke": it drifts, it wafts, its expands and seeps out of the boxes into which we attempt to allot it. More recently, philosopher Timothy Morton has suggested that contemporary ecological existence is one better conceived in relation to massive, distributed, non-linear phenomena or entities termed "hyperobjects" (including climate change, radioactive waste, even global finance) over and above "landscapes", which imply too much reliance on the detached surveying of calm, scrutinising observers. We cannot stand apart and survey these all-encompassing phenomena; they *are* the world(s) in which we inhabit (see Morton, 2013). Given the extensive upheavals and changes that seem to cascade across hierarchies of scale, transcend the individual and their perception, and unfold in impossibly complex ways, the conceptual value of a "landscape perspective" becomes unclear.

It certainly feels difficult to *place* contemporary crises, whether political, economic, environmental, or otherwise. Where exactly things are happening (or emerging from) and who (or what) is responsible are awkward questions to ask of a complex world. This is an age of churning seas filled with microplastics; extremist, nationalist politics that seem transnational in their rhetoric and influences; millions of social media interactions that pay little heed to spatial or temporal distances; and convoluted feedback loops and non-linear processes, in which we are all imbricated, driving ecological transformation. Nevertheless, studying cultural "landscapes" and how they manifest, change, or endure makes clear the many ways in which such crises (their experiencing, perception, knowing and response) *are* locatable. After all, how else is sea-level rise made tangible than via an encounter with a well-known coastal scene (see Matless, 2018)?

So, for the two of us writing this book, even amidst appreciation of phenomena that in many ways extend beyond ourselves and our immediate surroundings, there *is* value and relevance in landscape as a concept and an object of interest. This is not least because it directs attention towards "the middle ground" where, as environmental historian William Cronon (1996: 21) puts it, "we actually live". In response to the question of *how* to produce the "better knowledge" she seeks, Cuomo suggests we "engage, or recommit ourselves, to the project of getting closer". That is, a project which "involves attempts to bridge knowledge and action by bringing thinkers, knowers, and actors closer to the worlds affected by our actions and inactions" (Cuomo, 2003: 102). It strikes us, in its attention to a grounded and specific place, area or "scene" (as well as via its disciplinarity – see below) that landscape enables exactly this. It is not surprising, then, that "[l]andscape is increasingly being used to address larger questions" (Mason and Riding, 2023: 770), including those of the production and politics of knowledge, power, historic exclusion, and environmental transformation. Within this textbook, the terminology of cultural landscape freights a sense of holism that seems especially vital when working across disciplinary boundaries to address critical points of intersection or tension concerning the ways our world is changing.

Partiality: The Inclusiveness of Landscape

Attention to where the term landscape comes from, and the predominant ways in which it has been theorised and put to work, reveals its germination in contexts both Western and White (never mind male or elite – see Rose, 1993). Today's scholarship is undertaken amidst clear and widespread efforts to *decolonise* and pluralise the range of stories, knowledge-making practices, and voices within (and beyond) the academic conversation (see Bhambra et al, 2018). It is therefore obvious why some might query whether landscape is in fact the most appropriate framework for exploring human–environmental interactions. In other words, a certain anxiety concerning its origins and extant biases sees one minded to ask, "does this conceptual tool really work to tell a wide range of possible stories about human life?" (Bradley and Kearney, 2018: 290).

Ultimately, the answer to that question remains open and unresolved across this textbook. Landscape-oriented research is, by its very object of interest, *placed*, situated, and partial. It is emphatic in its appeals to, after philosopher Thomas Nagel, "the perspective of a particular person inside the world" (1986: 3). For us, the term *enables* forms of scholarship that house (and facilitate bringing together in conversation) a *diversity* of concepts, approaches, and methods. Even the most ardent adherent to one or other of the conceptual frameworks outlined within this textbook would be hard-pressed to claim other ways of thinking and doing landscape study were without value. In this way, landscape (like the concept of "place") *situates* interest, conjuring immediately the question of position(ality): who the viewer is, from where they are looking, and in what way are they engaged/immersed in their surroundings? It also enacts a pragmatic pluralism in its attempt to wrestle with the multiplicity of ways in which the environment is made and experienced (see Williams, 2014).

That is not to say the term flattens or resolves persistent hierarchies of knowledge. Nor can it simply shrug off its intellectual baggage. Its predominant association as the outcome of the interaction of "nature" and "culture" presupposes a dualist ontology far from universally held (Descola, 2013). We invite readers of this textbook to explore the limitations of landscape for themselves. Furthermore, we ask whether such limitations can be recast as *invitations*: to extend or critique existing debates, or even propose alternative forms of research practice. Seeing uncertainty and anxiety around landscape's partiality as *productive* in this way casts the negative edges of the concept (its limits) as a beckoning, even if that be towards the using of *different* concepts. For us, as authors, the term remains productive. Yet, we are mindful it may not always be so, as and when our areas of interest shift and change.

Unruliness: The (Non-)Definition of Landscape

A final set of apprehensions with regard to landscape lies in the assumption that it is unhelpfully broad or vague: simultaneously used to signify anything and nothing. In its unruliness and lack of definition, "landscape mocks scholars" (Stilgoe, 2018: 17). Just as each casual observer will perceive a scene differently based upon factors as diverse as their height, knowledge, or past experiences, so too will the landscape scholar utilise the term to convey an unruly diversity of meaning. This clamour of competing understandings leads John Stilgoe to argue that " 'landscape' is now a promiscuous word" (2018: 7). Whilst common, every day, ubiquitous in even a child's vocabulary, it has become so associated with other ideas, and multifaceted in its concerns, that questions of its utility and comprehension must be asked.

There are, however, benefits to such vernacular usage. In terms of scholarship, the wide remit of landscape brings together people from different disciplines and puts them into conversation with one another. As Barbara Bender (2002: S106) writes, "[l]andscapes refuse to be disciplined. They make a mockery of the oppositions that we create between time (History) and space (Geography), or between (Science) and culture (Social Anthropology)". In being thus, however, landscape helps us to meet other disciplinary approaches "halfway" to potentially cultivate "joint modes of thought" and "points of contact" (Reinert, 2018: 506). The amorphous conceptual space that landscape occupies can, therefore, be a positive trait which, if embraced, holds the scholar's attention open to a wider range of phenomena, ideas, debates and approaches as it affords an engagement with them.

It is also worth recognising that, while the term may indeed be promiscuous, it is everywhere in policy, management plans, legal frameworks, and NGO proposals that have real-world implications. It is a term in use, regardless of its problems, that we deploy to describe everything from the availability of financial resources ("the funding landscape") to the designation of areas for special conservation or heritage value. Whether or not we agree what it means, "landscape" has force and presence. Consequently, as Friess and Jazeel (2017) argue, it is increasingly important that we reckon with the ways in which landscape is (or can be) conceptualised, and how such understandings may underlie, or could be applied within, a range of different fields or in support of different vested interests. Given such wide deployment, moves to disengage with landscape on the grounds of its unruliness would appear at best naive. Like other cultural "keywords" (see Williams, 1976), occupying a broad remit and link into myriad-associated terms and ideas, landscape is not neutral. Exploring the complexity of its meanings and applications, especially as they conflict or contradict each other, is vital.

Ultimately, we remain mindful that landscape is a term with limits, weaknesses, and baked-in assumptions. A sprawling notion with this kind of baggage demands *critical* attention unafraid to question its relevance, partiality, and lack of coherence. Nevertheless, we find in landscape scholarship, past and present, a hopefulness about the value of this concept. It is that optimism which motivates the production of this textbook. Whether you share that view, and whatever it may be that the ideas and approaches within this textbook allow you to do, remains to be seen. In the oft-quoted terms of philosopher Brian Massumi, landscape, like any concept, "is a brick". As such, "[i]t can be used to build a courthouse of reason. Or it can be thrown through the window" (Massumi, 1987: xi).

Positionality

Any scholar, as the editors of the multidisciplinary journal *Landscape Research* emphasise, writing or working on/within cultural landscapes is always already "situated in landscapes" (real and imagined). These "provide the vantage points from which we look at language and knowledge" and inform the weaving of "a tapestry of experiences" that both link us with, and shaping our perspective towards, the world we inhabit (Waterton et al, 2022: 2). Indeed, since "any landscape is composed not only of what lies before our eyes but what lies within our heads", it could even be that, faced with the same scene, "we will not – we cannot – see the same landscape" as another person (Meinig, 1979: 33). However possible it may be to reckon with our own subjective experiences – never mind empathise or appreciate those of others – clearly the issue of a researcher's *positionality* matters to the kinds of cultural landscape accounts (and cultural landscapes) they engage with, respond to, and (re)produce. By positionality, here, we mean the sense in which we (the authors) and you (the reader) are " 'positioned' within social, cultural, and economic dynamics of identity in ways that effect how we engage with, interpret and produce knowledge about the world" (Sanyal, 2020: 26).

Positionality concerns not only how research is designed, carried out, and written up. It also includes how (and what) we read and the broader question of "what we see as legitimate knowledge" (Sanyal, 2020: 26). To begin this textbook in the spirit of open, self-reflection, and encouraging a critical reading on the part of those engaging its content, we consider our own positionality as authors with particular attention to scholarly lineage and disciplinary assumptions; and our lived experiences of cultural landscapes. As with the discussion of anxieties above, the aim is not to close-down or pre-empt critique. Indeed, it is the very opposite: an effort to illuminate gaps, limits, and spaces in between the points we make, the works we cite, and the lines of thought we trace as invitations for others to follow.

Our Arrival into Landscape

As geographers whose backgrounds have been shaped by geographical disciplinary norms, debates, and concepts; and no matter how widely we might range in our writing or reading, our ways of understanding and practising landscape occur within that knowledge context. Indeed, both of us owe substantial portions of our undergraduate (and postgraduate) education, our formative exposure to ideas, concepts, debates, and styles of researching, to individuals who themselves have been positioned in relation to discussions of "landscape" in not insignificant ways (including within this textbook). We could even go as far as to say that many of those scholars working on landscape whom we encountered during our studies inspired us to the writing of this textbook, which it would be awkward to pen without some discussion of their work.

While accepting our formative conditioning, we do attempt to get outside the ways in which cultural landscapes have been discussed in geography and consider its resonance or debate in cognate fields across the chapters in this book. Nevertheless, and despite our efforts to expand how landscape is articulated in our own minds as much as those of the reader, we acknowledge that the traces of our geographical training are probably rather evident to a non-geography-based scholar. We encourage the reader to be aware of our background and to feel free to dissent from it, and the effects of how we set out conceptual approaches to landscape and culture in terms of their reproduction of quite specific disciplinary norms. We also assert that there is a risk that we, as geographers, end up *thinking for* rather than thinking from these other perspectives that constitute the diverse, "populated worlds" of landscape studies (see Puig de la Bellacasa, 2012). We therefore suggest that those interested in adding depth to disciplinary approaches to cultural landscape studies do so by way of direct engagement with those works cited (and many more besides).

Our Experience of Landscape

It is also important to acknowledge that our experiences of cultural landscapes, and the places we tend to think through and with, are primarily British. We are white and middle class and have families with small children. We present these facets of ourselves not to demand that everything we have written be read through them, but to encourage reflection on how a different author (or authors) writing of, or with, different histories and geographies, never mind social identities, might have produced a very different version of this book.

As we critically discuss within many of the chapters that follow, the term "landscape" has an often rural, nonhuman bias in its associations. It tends to reproduce a discourse of nature as existing separate and subordinate to an objective viewer or to lead us into romantic, traditional, visions of the past. We are perhaps complicit in this association too, being as we share passions for hillwalking, mountains, and the Highlands of Scotland; no doubt because these kinds of activities and places have figured centrally to our upbringings, continuing lives, and both academic and non-academic interests. There is a tendency for discussion to wander into the hills in terms of the examples we reach for to illustrate the ideas or approaches that we consider. Whilst we do make every effort to engage urban and suburban scenes (not to mention a variety of other "-scapes") such selection bias is worth acknowledging. This is not least because recognising that the cultural landscapes we encounter (in living or reading/writing) are "ideologically loaded as a starting point rather than an end point" (Hodson, 2022: 281) prompts us to face up to the ways in which they, even implicitly, serve certain normative interests and agendas at the expense of others.

Chapter Overview

This book is divided into 13 chapters which chart various approaches to the study of cultural landscapes. These range from those approaches which consider the landscape as a "finished" object or document, to be interpreted and read for the traces or values of its parent culture (Chapters 1, 2, 4); through chapters that engage with the processes by which landscapes are (re-)produced in the context of relations of power, capital, and in relation to wider norms or conventions (Chapters 3, 5–7); to those considering landscape in the context of relational, embodied, affective and more-than-human understandings of culture, experience and perception (Chapters 8–11). Throughout the textbook, examples and activities (see "Key Pedagogical Features") provoke consideration of the ways in which these approaches can be applied to studies of cultural landscapes or else inform understandings and engagements with it.

Chapter 1 concerns, broadly, what it means to consider the study of human cultural–environmental relations via a "landscape perspective". In doing so, it charts the origins of interest in Anglophone cultural landscape study and particularly the influential role of Carl Sauer's morphology in carving out a more open-ended, rather than deterministic, appreciation of the diversity of cultural landscapes. Whilst many might be tempted to relegate Sauer's propositions to the history of landscape scholarship, we try to trace some of the ways in which his ideas landscape persists to inform a contemporary landscape perspective.

The ways in which cultural landscapes are considered in relation to representations of the environment, particularly the idea of landscape art, informs Chapter 2. Here we flesh out a sense of landscape as socially constructed through particular "ways of seeing" (shared visual conventions) that define how, for what purposes, and in which contexts the environment is judged as a "landscape". The chapter also considers the ways in which acts of seeing/representing landscape are situated within a particular subject position. In turn, it follows recent attention to the *doing* of creative work can inform an understanding of cultural landscapes, considering representational acts as themselves taking place within the world (linking to later discussions in the textbook).

Chapter 3 draws from landscape political ecology and critical Marxist scholarship to emphasise the importance of examining the historical formation of scenes by way of inequality, struggle, and the relations defining of economic modes of production. Emphasising the need to "materialise" our sense of the cultural landscape and attend to its manifestation and enacting of "work" in service to the economy, the chapter considers how different landscapes can help us conceptualise the relationship between environment and labour processes.

Continuing some of the implicit consideration across previous chapters of the ways in which cultural landscapes can be "read", Chapter 4 conceptualises landscapes as texts. In doing so, it elaborates the methods by which meaning can be gleaned from the close study of the values, meanings, and narratives "written into" the landscape (by cultural groups for different purposes). There is also, however, the matter of how such readings, in the widespread adoption of post-structural theories of textuality, are inherently multiple, unstable, and continuously liable to reworking.

Such themes are developed in Chapter 5 which engages explicitly with the discursive reproduction of the environment in a cultural context. The notion of the "moral landscape" is used to exemplify the ways in which our interactions with landscape are shaped and regulated in association with institutions, norms, and the performance of subjectivities, cultivated or suppressed depending on their positioning within a wider set of power relations. Thus, the morality of landscape displays how cultural landscapes, as discursive, serves to encourage or discourage certain kinds of environmental subjects, and land uses.

The discourses of cultural landscapes, and their production through struggled-over histories of inequality, representation in cultural texts, and inscription with the values of the powerful, are encountered in the specific context of colonialism and imperialism in Chapters 6 and 7. The former considers the cultural landscape as a manifold of imperialist relations between coloniser, Indigenous and colonised peoples, and the environment, as well as the ways in which the cultural and material processes of imperialism served to produce distinct formations of landscape. The latter chapter considers what it means to engage with cultural landscape scholarship (and the concept of landscape itself) in the wake of this history. Considering multiple avenues for the "decolonisation" of landscape, conceptually and materially, Chapter 7 questions whether this term can be deployed at all and what it means to conceive of landscape after Empire and in relation to alternative, Indigenous cosmologies, and ecologies.

As discussed above, the remaining chapters draw much resource from what have been termed non-representational or more-than-representational theories and approaches. Chapter 8 engages the long tradition of landscape phenomenology and the conception of the cultural landscape as something emergent from the perception of the environment by a sensing body that is always, already immersed within it. In turn, the chapter also examines some of the applications and critiques of the phenomenological approach within cultural landscape studies, as well as more recent articulations of a "post-phenomenology" that tries to incorporate a sense of the subject's cultural context (as raced or gendered, for example); as well as the role of other entities or agencies in changing (or producing) the capacities for body-world encounters.

Chapter 9 explores the many ways in which the "feeling" of landscape, as something that prompts an emotional or otherwise visceral response escaping articulation in language, can be theorised and understood. It engages with feminist scholarship on emotion, as well as the concept of affect, to examine how the landscape comes to be changed or charged through experiences of it, and how it impresses itself upon us, in ways that might escape or come prior to cognition. The tensions between affect and emotion, and the way each term can be usefully spatialised for the study of cultural landscapes are explored, with particular emphasis on the idea of the "affective region".

Recognising that the cultural landscape is a lively field composed of various kinds of material forces, processes and agencies that press upon or interact with the subject opens consideration of its "more-than-human" character – the focus of Chapter 10. The ways in which attention to animals can be incorporated into, and complementing (or complicating) of the approaches to the landscape already discussed elsewhere in the textbook is considered. The chapter also looks at the ways in which acknowledging the presence of other "actors" in the landscape (nonhumans of all kinds) demands a more *relational* conception of culture and the environment, and a consideration of the ability to "make a difference" as being more dispersed across space and time.

Whilst most of the chapters in the textbook have considered cultural landscapes, broadly, in terms of humans (and others) being present in, or presented with the content of, the environment in different ways, Chapter 11 takes instead the question of *absence* as its central concern. Connecting to broader questions concerning the spectral, ghostly, and negative across the humanities and social sciences, the chapter uses the notion of haunting to explore how it is that cultural landscapes come to be defined or experienced in relation to that which is not (or no longer) there. The role of memory, trauma, and time in shaping an encounter with cultural landscapes of absence is elaborated, as is the politics of absence, which ghosts are manifested (or not), and the non-representational ways in which these spectres are made knowable.

The penultimate Chapter 12, and the last to consider a particular conceptual approach to culture and landscape, examines the notion of therapeutic landscapes. The relationship between environment and health, conceived of via landscape, is explored, including the variety of approaches by which that connection has been examined or interpreted. As such, the chapter is more driven by the practical questions of how the interlinking of health and environment can be managed and informing of landscape policy and management, than with a specific theoretical angle on "culture". Nevertheless, many of the approaches of previous chapters feature or inform the work under discussion here.

Closing the textbook, Chapter 13 turns to matters of method and the active study of cultural landscapes by students, researchers, and teachers. It considers a variety of methodological themes, cross-cutting the diverse ways into the cultural landscape that the textbook has showcased, to provide suggestions and inspiration for how thinking *about* landscape can inform the practical investigation *of* landscapes. Reflecting in part contemporary questions surrounding the ethics of fieldwork in an age of climate crisis, forms of research that involve both "being in" cultural landscapes and not are presented for the reader to take forward as desired.

Key Pedagogical Features Explained

Throughout each chapter, we utilise several key features to assist the reader in their journey into studying cultural landscapes. These include the following: the identification of relevant key concepts; content boxes; pauses for thought; and suggested seminar activities and readings (for both learners and tutors).

Key concepts: Each chapter identifies up to five key concepts relevant to their concerns. We do not offer a singular definition of these concepts for you, though they are often defined (or given definition "in use") during each chapter's discussion. Rather you are encouraged to note them down, pay attention to how they appear, consider what they may mean (practically, conceptually, contextually), and follow them up in terms of how they are critically applied or understood by other scholars.

Content boxes and "pauses for thought": Chapters make use of content boxes to introduce materials (an illustrative example, a discussion of an area of scholarship, or key critiques of important ideas or approaches) that are additional to the main text, yet expand upon or support understanding of the chapter's central concerns. "Pauses for thought" encourage the reader to take a break and consider where examples from their own life, experience, knowledge, or wider reading might be usefully connected to the arguments under discussion. It is often through making these kinds of connections that ideas resonate, and so we aim to foster that practice here.

Suggested seminar activities and readings: These are intended as more practical or discursive-based activities that provide an alternative way into, or a chance to complicate or elaborate, some of the key ideas from the chapter. Those reading the textbook as independent learners might find these activities provide inspiration for how to mobilise different approaches to landscape and practical suggestions for extending knowledge. For tutors, these activities could be used or repurposed in the design of seminars for courses they teach. Likewise, recommended readings comprise examples of key articles that we (the authors) have found helpful in setting out, applying, or critiquing the topics discussed in each chapter.

References

About YSP. n.d. *Yorkshire sculpture park website*. Available at: https://ysp.org.uk/about-ysp. Accessed on 28 August 2024.

Bender, B. 2002. Time and landscape. *Current Anthropology* 43(S4): S103–S112.

Bhambra, G., Gebrial, D., and Nişancloğlu, K. Eds. 2018. *Decolonising the University*. London: Pluto Press.

Bradley, J.J. and Kearney, A. 2018. 'You whitefellas pull it all apart': Epistemic learnings in exploring landscape. In: Howard, P., Thompson, I., Waterton, E. and Atha, M. Eds. *The Routledge Companion to Landscape Studies*. Routledge: London, pp. 288–298.

Cronon, W. 1996. The trouble with wilderness or, getting back to the wrong nature. *Environmental History* 1(1): 7–28.

Cuomo, C. 2003. *The Philosopher Queen: Feminist Essays on War, Love, and Knowledge.* Oxford: Rowman & Littlefield.

Descola, P. 2013. *Beyond Nature and Culture.* Chicago, IL: University of Chicago Press.

Friess, D.A. and Jazeel, T., 2017. Unlearning "landscape". *Annals of the American Association of Geographers* 107(1): 14-21.

Garlick, B. and Hunt, R. 2022. Landscape. In: Ballamingie,P. and Szanto, D. Eds. *Showing Theory to Know Theory: Understanding social science concepts through illustrative vignettes*. Vol. 1. Ottowa, ON: Showing Theory Press, pp.240-249.

Hodson, L. 'Picture perfect' landscape stories: Normative narratives and authorised discourse. *Landscape Research* 47(2): 271–284.

Landscape. 2024. *Cambridge dictionary online*. Available at: https://dictionary.cambridge.org/dictionary/english/landscape. Accessed on 28 August 2024.

Mason, O. and Riding, J. 2023. Reimagining landscape: Materiality, decoloniality, and creativity. *Progress in Human Geography* 47(6): 769–789.

Massumi, B. 1987. Translator's Foreword: Pleasures of Philosophy. In: Deleuze, G. and Guattari, F. Eds. 2013. *A Thousand Plateaus.* London: Bloomsbury, pp.vii–xiv.

Matless, D. 2018. The anthroposcenic: Landscape in the anthroposcene. *British Art Studies* 10. https://doi.org/10.17658/issn.2058-5462/issue-10/dmatless

Meinig, D. 1979. The beholding eye: Ten versions of the same scene. In: Meinig, D. Ed. *The Interpretation of Ordinary Landscapes.* New York: Oxford University Press, pp. 33–48.

Morton, T. 2013. *Hyperobjects*. Minneapolis, MN: University of Minnesota Press.

Nagel, T. 1986. *The View from Nowhere.* Oxford: Oxford University Press.

Puig de la Bellacasa, Maria. 2012. 'Nothing Comes Without Its World': Thinking with care. *The Sociological Review* 60(2): 197–216.

Reinert, H. 2018. The haunting cliffs: Some notes on silence. *Parallax* 24(4): 501–512.

Reitter, P. and Wellmon, C. 2021. *Permanent Crisis: The Humanities in a Disenchanted Age.* Chicago, IL: University of Chicago Press.

Rose, G. 1993. *Feminism and Geography.* Cambridge: Polity Press.

Sanyal, R. 2020. Positionality. In: Wilson, H. and Darling, J. Eds. *Research Ethics for Human Geography.* London: Sage, pp. 25–32.

Stilgoe, J. 2018. *What is landscape?* Cambridge, MA: MIT Press.

Waterton, E., Castán Broto, V., Fairclough, G., Jones, K., and Watt, L.A. 2022. Editorial: Thinking and wwriting with care. *Landscape Research* 47(1): 1–9.

Williams, D. November 2014. Making sense of 'place': Reflections on pluralism and positionality in place research. *Landscape and Urban Planning* 131: 74–82.

Williams, R. 1976. *Keywords*. Oxford: Oxford University Press.

1 A Landscape Perspective

Introduction: Landscape as Ontology

Opening a special issue of *Landscape Research* in 2023, interdisciplinary urban researchers Vanesa Castán Broto and Enora Robin outline their aim of situating studies of "infrastructure" within social context. As they argue, too often infrastructures – incorporating roads, public transport configurations, energy networks, and sanitation systems – are analysed and discussed in ways that detach these (albeit large) "objects" from the social and ecological *worlds* in which they are immersed and with which they interact. The result is that infrastructures are taken for granted as stubborn, unchanging, or permanent. The processes that make them seem so, and the many ways in which infrastructure *is* often changing over time, get ignored. One way of countering this, they suggest, is the adoption of "a landscape perspective". For these editors, "landscape thinking" (Castán Broto and Robin, 2023: 167) provides a way of engaging through direct attention to the variously observable – or empirical – interactions between infrastructural systems and their wider environments. Showcasing this "landscape perspective", their special issue gathers contributions documenting different urban sites and case studies from researchers working across several different academic disciplines.

We open with this brief example because it highlights how "landscape" can be understood to name a particular *ontology*. In other words, the term landscape encapsulates a way of understanding what the world is (and something of how it works) that serves as a starting point for subsequent research, be that about infrastructure or something else. Whilst landscape appears as a very broad term, concerned with "everything and anything" (Castán Broto and Robin, 2023: 168), it is also clear that, for many, it provides a useful spatial framework. This is not least because it places, on the one hand, "society" or "culture" into firm and inescapable dialogue with, on the other hand, those things termed "nature" (the substance and materiality of the environments in and through which socio-cultural activities occur) in a particular area. What this chapter seeks to do is examine some of the origins and key foundations for this way of thinking landscape, first and foremost, *as a framing term* that structures research into a range of phenomena across the social sciences, humanities and beyond (for example ecology).

DOI: 10.4324/9781003172451-2

We also use this chapter to argue that a sense of landscape as a holistic "unit" of space with particular, unique qualities remains central to much work today. A key figure underpinning this way of thinking is Carl Sauer, whose work significantly proposed the "cultural landscape" as an object for scrutiny. A familiar character in many introductory lectures (and textbook chapters) concerning Euro-American traditions of landscape, it can feel, as John Wylie (2007) observes, that any retrospective discussion of landscape is *beholden* to mentioning Sauer. We hope to rejuvenate consideration of his ideas here by emphasising that, despite many valid critiques, Sauer's central proposition – that landscape provides the "unit concept" for examinations of human–environmental interaction – continues to hold across a good deal of contemporary and diverse writing. Our aim is *not* to position Sauer (or others) within a definitive "family tree" of landscape. Rather, we want to suggest lines of affiliation or resonance as we reflect on how many scholars continue to use landscape in their empirical work. As we also consider, the ontology of landscape as empirical fact and material context is also prevalent in another way in contemporary (geographical) scholarship – that is in the midst of a (re)turn to place writing as method and the championing of "geographical portraiture" (after Lorimer, 2019) as an analytic device that deploys a landscape perspective in the guise of creative nonfiction.

Chapter Outline

Introducing a "landscape perspective", this chapter explores the emergence of landscape as a term referring to the environment as an object of human interest and interaction. It charts the shift from understandings of landscape via environmental determinism to the anti-determinist "cultural landscape" concept proposed by Carl Sauer, drawing on ideas of *landschaft* to examine culture as a key component of the physical and social processes producing a unique "morphology" of landscape. The chapter highlights the key critiques of this morphological approach, including its "superorganic" treatment of culture, foreshadowing the concerns of subsequent chapters. In concluding, we suggest the continuing relevance of this kind of landscape perspective, making connections between Sauer's morphology and contemporary landscape studies in general.

The Emergence of a Landscape Perspective

In terms of etymology (the study of the origins of words and their changing meaning), the term "landscape" is often understood to originate in the Middle Dutch language of the sixteenth century, combining *land* – "land" – and *scap* – equivalent to the suffix "-ship". Here "ship" operates as in words like "ownership", "kinship", or "friendship", to broadly indicate a state of being or form of relation associated with a corresponding root word (e.g. the state of friendship associated with friends or the state of owning something as ownership). Thus, *landscap* – anglicised as landscape – was a term deployed with reference to particular forms of art that focused on environmental surroundings or scenery (an association persisting today – see Chapter 2). In the nineteenth century, however, the term would move outside of this narrow, aesthetic usage and be applied in ways intended to characterise or refer to the shape of the *material* land, as well as those social and political structures organising life upon it. In this sense, landscape would approach and become entangled with the origins and associations of the German language term *landschaft* (much discussed by Olwig, 1996 – see Chapter 3). As attended to later in this textbook, appreciating these specifically European origins of "landscape" can prompt the question of whether it can (or should) be transposed into or applied onto other cultural contexts (see Chapter 7).

An early and influential application of the term landscape as a spatial unit encapsulating a sense of *holism* (i.e. the treating of the environment as a whole, constituted from a variety of intersecting and diverse material processes and parts, with its own resulting properties) can be found in the writings of naturalist and proto-geographer Alexander von Humboldt (see Minca, 2007). Significant for his contributions to thinking spatially about the distribution of vegetation in relation to climate, alongside pioneering and creative ways of *representing* such relationships via maps and charts, von Humboldt conveyed a notion of "regional diversity" as something emerging from and reflected in the geographical variations in biophysical processes. The term "landscape", therefore, enabled him to capture in a single term the diverse array of environments he documented, suggesting that each was a kind of "whole" made up of its various physical processes and phenomena (see Antrop, 2020).

Over the course of the nineteenth century, this holistic sense of landscape was expanded to include humans and their activities amongst the many processes producing different environments. At first, this relationship was understood in broadly deterministic ways. Latterly an appreciation of the possibilities that landscapes afforded for human development took hold. That is, attention shifted from considering how the features of landscapes *directed* or influenced the societies they contained to instead appreciating landscapes as *outcomes* of various possible forms of living, as we now discuss.

From Determinism to Possibilism

In the mid-nineteenth century, naturalists Charles Darwin and Alfred Russell Wallace advanced ideas about natural selection, environmental influence, and the emergence of specialised lifeforms through evolution, based upon observations accumulated during expeditions to the Pacific and Southern America in the early-mid-nineteenth century. These ideas were subsequently then set in relation to *social* phenomena and *human* societies, reinforcing an outlook known as environmental determinism. This idea – that environmental conditions could explain human social and cultural differences – was not new to Western thought (see Descola, 2013). For example, the ancient Greeks had theorised that the earth was divided into a number of climatic zones, with each manifesting different physical and cultural expressions of humanity, and, in the Middle Ages, such ideas were taken up again to inform a variety of ideas about the monstrous savages to be found beyond Europe's borders (Sharp, 2022 – see Chapter 6). In the nineteenth century, and in the context of both the British Empire and longer debates concerning the environment's role in shaping human development, Darwin's theories of natural selection were not confined in their influence to studies of natural history. They took hold in a wider social context to explain, naturalise, and justify *human* social, cultural, political, and economic activities (like industrial capitalism), as well as the dominance of certain societies over others (see Hinchliffe, 2007). Consequently, productive, advanced, or "civilised" societies were viewed as those that had developed in situations with favourable environmental conditions.

The subsequent emergence of "anthropo-geography" in the work of figures like German geographer Friedrich Ratzel demonstrates how such biological theories informed understandings of human society, development, and spatial behaviour (Farinelli, 2000). Ratzel's influential (and subsequently devastating) concept of *Lebensraum* ("living space") was an attempt to conceptualise how humans (and other creatures) lived in their environments via consideration of the relationships between life and space. The latter, a material domain, presented certain limits or opportunities for different lifeforms. Thus, "the evolution of life [was] a process of summation" of those forces acting upon it – in short, of natural selection – conceived as specific "spatial conditions" (Ratzel, 2018 [1901]: 61). For Ratzel, "the manifestation of life" required a "mastering of space" as different lifeforms organised themselves in response to environmental characteristics in order to flourish (2018: 63). *What* you were became figured as a question of *where* you were.

Ratzel and others, such as American geographer and writer Ellen Churchill Semple (1863–1962), thus applied ideas of natural selection to advance the idea that human societies and cultures were also an outcome of environmental conditions and pressures. As Semple wrote in the introduction to *Influences of the Geographical Environment*, "man [sic.] is a product of the earth's surface" (1911: 1). Human differences (for example racial traits and ethnic diversity) were understood to be *caused* by environmental differences. Cultural variation was understood as a function of ecological conditions (Winchester et al, 2013). These ideas circulated widely, informing, infusing with, and promoting a pernicious brand of scientific racism. Under British and wider European colonialism, such assumptions were embedded in the thinking and practices of geography, anthropology, medicine, and governance during the nineteenth century (Keighren, 2006). The consequences could be contradictory: Non-European cultures might be cast as inherently inferior, reflecting supposedly diseased, degraded, or in some way *lesser* environmental situations as evidence within discourses of "tropicality". Simultaneously, understanding the potentially damaging influence of the environment prompted a "civilising mission" that saw non-Western landscapes (and peoples) subject to forms of *improvement* (see Chapter 6). However it was applied, then, the basic notion of environmental determinism often framed how the relationship between land and life was to be comprehended until the early-to-mid-twentieth century.

An alternative perspective, building on the work of a number of significant anthropologists and geographers working towards the end of the nineteenth century, emerged and formed the foundations for the study of cultural landscapes. Whilst figures such as Ratzel (and before him, Carl Ritter) had considered the *Kulturlandschaft* (cultural landscape) – landscape "modified" through human activities – it was as a natural and determined outcome of environmental pressures. Developing this notion subsequently, however, scholars like German geographer Otto Schlüter argued against a deterministic framing of cultural landscapes to instead assert that they might yield an understanding of the "formative influence" of humans upon their surroundings' historical transformation in relation to different interests, needs, or values (see Jones, 2003).

Working in the latter-nineteenth century, geographer-turned-anthropologist Franz Boas would influentially champion attention to culture and cultural change that resisted the "evolutionary" understanding of its development. Such work, he felt, too often based itself on studies of museum collections, rather than *field* work that engaged with living societies and their cultural practices to understand how their ways of existing in the environment developed and shaped its formation over history. Boas – and those he influenced, including protégé Alfred Kroeber – proposed that culture required examination in *context.* They understood cultural existence (comprising practices, rituals, artefacts, languages, myths) as a separate domain of reality, apart from the organic domain of material existence. A "superorganic" phenomenon, culture was a force above or apart from both the processes of nature or the everyday business of social relations (a proposition that would come under significant critique – see below) and could not be explained as merely the "adaptation to environment". For Kroeber, "the explanation of culture [lay] ultimately in culture itself". That is, an analysis of symbols, beliefs, value systems, and ideas or concepts (themselves often attached to material artefacts or present in various forms that could be "salvaged" via fieldwork) could yield an insight into what Sauer would later term the "inclinations" of human communities towards this or that aspect of their environment and the culture that flowed through them (see Solot, 1986: 513–514).

Kroeber joined the University of California at Berkeley in the early part of the twentieth century, bringing Boas' anti-deterministic ideas and applying them to his own work with (or on) Indigenous Californian societies (Jacknis, 2002). Whilst certainly steeped in colonial ideas of knowledge production and associated power relations, his work laid the ground for a more "possibilist" (rather than determinist) consideration of environment–culture interactions. Such an outlook appreciated the role of human culture as a force *shaping* the environment rather than being determined *by* it. Indeed, a broader interest was developing at this time towards the characterising and study of cultural character at the regional scale. In the terms of French geographer Paul Vidal de La Blache (1911), the aim was to discern how particular "ways of life" (*genre de vie*) left their imprint upon places. Into this context came Carl Sauer.

The Morphology of Landscape

Born 1889 in Missouri, USA, Carl Sauer would study for a PhD at the University of Chicago (1909–1915) before finding work afterwards at the University of Michigan. As Kent Mathewson (2018: 10) argues, from his doctorate onwards he was "at odds with mainstream American geography" and its dominant outlook of "cause-and-effect geographical determinism". In 1923, he moved to Berkeley and became, at 33, Chair of Geography. Working with and influenced by the ideas of the anthropologist Kroeber, as well as contemporaries in Europe and the concept of "landschaft" they deployed (see Arntz, 1999; Olwig, 1996), Sauer elaborated his own, anti-deterministic approach to landscape and the study of its cultural "morphology" that would later be characterised as the "Berkeley School" approach to landscape studies. Its influence was widespread, not least for "discrediting and perhaps even derailing" environmental determinism as the primary conceptual framework for explaining human–landscape interactions (Mathewson, 2018: 11).

In his 1925 essay, "The Morphology of Landscape", Sauer outlined how the "cultural landscape" emerged as the material consequence of human and non-human *interactions* unfolding over the course of time. Sauer envisaged the task of landscape study as "the establishment of a critical system which embraces the phenomenology of landscape [how it is encountered, experienced, and perceived by an observer], in order to grasp in all of its meaning and colour the varied terrestrial scene" (Sauer, 1925: 320). He did not wish to begin with an overarching theory to explain the landscape (such as the notion that environmental influences determined cultural development). Rather, Sauer argued that landscape researchers build understanding of how specific environs came to be formed from the ground up by way of direct fieldwork. Landscape named, for Sauer, "the unit concept" of his own discipline of geography, constituting "the peculiarly geographical association of facts" that characterised a portion of the earth's surface (Sauer, 1925: 321). Crucially, the landscape one sought to account for was "not simply an actual scene viewed by an observer" but "a generalization" of a place type, "derived from the observation of individual scenes" and clarified through comparative observation (Sauer, 1925: 322).

Central to Sauer's morphology was that landscapes emerge from the *interaction* of human and "natural" elements. There existed, to begin with, the "natural landscape": the environment "before the introduction of man's [sic.] activity" (Sauer, 1925: 333), expressing the forces of geology, climate, and vegetation, as well as their interaction over time. This natural landscape was to be appreciated *holistically* in terms of how each of its aspects interrelated to produce its observable form. Over time, natural landscapes were transformed into specific "cultural landscapes", incorporating of "the forms that man [sic.] has introduced" (Sauer, 1925: 333). Thus, a cultural landscape (or succession of overlain cultural landscapes) comprised the various ways in which human society had altered their surroundings in tangible, *material* ways: i.e. "man's [sic.] record upon the landscape" (Sauer, 1925: 342). Importantly, then, Sauer was *not* interested in "the energy, customs, or beliefs of man [sic.]" (1925: 342) but rather their physical traces – a stance subject to much critical discussion in later landscape study.

Thus, to summarise the morphological approach:

> Culture is the agent, the natural area is the medium, the cultural landscape is the result.
>
> (Sauer, 1925: 343)

Sauer therefore turned environmental determinism on its head: culture named the "footprint" of human activity and society left upon a malleable earth (Winchester et al, 2013). In turn, landscape studies following the Berkeley School model often focused on mapping the distinctive material character ("morphology") of given landscapes, as well as tracing the spread or diffusion of cultural practices by virtue of traces left in different environments (for example building methods or styles, or farming techniques). This latter question, reflecting the centrality of Sauer's interest in the relationship between *time* and landscape and his treatment of cultural change in terms of attention to cultural objects (rather than processes), saw him "sidestep" the matter of *how* cultures developed. Instead, Sauer sought to trace a "sequence of forms", or changing cultural areas, over history and across space, to produce a genetic map of cultural change and diffusion (Solot, 1986: 516–518).

Sauer's interest in time and culture was developed in Europe by others, such as Torsten Hägerstrand and his approach to "time geography", that incorporated an ecological account of landscape (see Box 1.1). Sauer's morphology also found an analogue in the emergent *regional* landscape studies that featured in much geographical scholarship prior to the Second World War by figures such as Dudley Stamp, Eva Taylor, E.W. Gilbert, and H.C. Darby. Attempting to holistically understand and historicise specific regional forms, such work fundamentally sought to describe and synthesise the sense of regional character or identity as a "unity" of human and nonhuman processes and relations (see Paasi, 2020). Indeed, in Sauer's later work, a sense of "regionality" (see also Chapter 9) was discussed as a "personality": "the whole dynamic relation of life and land" that it uniquely expressed (Sauer, 1941: 354).

Box 1.1 Torsten Hägerstrand on Landscape as an "Integrative Concept"

Swedish geographer Torsten Hägerstrand (1916–2004) is commonly associated with efforts to describe behavioural and spatial relationships during the height of the post-war "quantitative revolution", when many social sciences sought to affirm their scientific status by embracing statistical methods and mathematical modelling. Key contributions include his examinations of intellectual "diffusion" (or spread) of cultural ideas and practices, as in agriculture; his studies of travel times as shaped by transport networks and the tracing of individuals' "space-time paths" through their environment over a specified period ("time geographies"); and contributions to regional planning and governance in his native country. Like many quantitative social scientists, Hägerstrand sought to outline abstract laws explaining or predicting spatial forms and behaviours deductively, rather than test or construct such models via real-world study and data collection (see Flowerdew, 2011).

Recent reappraisals of his work and later Swedish-language output have emphasised the importance of Hägerstrand's landscape perspective in geography. He became deeply interested in landscape as a concept offering "a language to describe the interplay between society and nature, and to study a wide range of (abiotic, biotic, or societal) phenomena on equal terms" (Qvistrom and Wästfelt, 2020: 683). Thus, Hägerstrand in Sweden, like Sauer in America, saw landscape as a useful, interdisciplinary framework or "integrative concept" (Qvistrom and Wästfelt, 2020: 683) to grasp specific and complex relations between "nature" and "culture" as they defined places humans lived. Landscape, for Hägerstrand, both helped convey an interest in more sustainable development but also evoked something of the lived, everyday material world, encompassed in terms of a human-nonhuman–environmental ecological whole, or "geographic now" (Hägerstrand, 1983: 239 – cited in Qvistrom and Wästfelt, 2020), that was constantly in process. Recent scholars, therefore, have reappraised and celebrated Hägerstrand for his landscape thinking and embracing of a landscape perspective.

For geographer Daniel Gade, reviewing the legacy of the Berkeley School, Sauer was a *romantic* thinker of landscape. This, he says, is not meant in the sense of what the word usually implies: being irrational (or anti-rational) and tending to reject pragmatism or realism. Rather, it is to acknowledge his leanings towards the rural; his resistance to overarching narratives or theories (rejecting Marxist scholarship as much as classical or reductionist models) that might "[impose] uniformities of structure on human life"; and his "strong anti-utilitarian bias". As Gade summarises:

> In Sauer's way of thinking, there is no structure of things, no universal rules, and no pattern to which one must adapt, thus accounting for his refusal to promote divisions of knowledge or to engage in epistemological debate about the boundaries of geography. Like all romantics, he had a historicist perspective and a strong conviction that we cannot really look forward without looking back. Sauer opposed any canonic approach to method or technique, considering them to be iron cages constraining his imagination. His idealization of diversity and the notion of pluralistic worlds, his exaltation of the local and contingent, and his interest in the uniqueness and individuality of objects in their concrete totality

were all part of his romantic consciousness. The intuitive, the continuous, the organic, the disordered, and the traditional defined who Sauer was.

(Gade, 2014: 118)

(Re)Evaluating Sauer's Landscape Approach

Sauer's model of cultural landscape had an impact within landscape studies that is "hard to exaggerate" (Mitchell, 2000: 20). It is, however, an approach not without substantial critique. Indeed, discussions of the limitations to the Berkeley School approach have, in turn, fed into the development of numerous *other* ways of "doing landscape" elaborated across this textbook. We briefly elaborate on some of these critiques here, mindful of their being treated in more detail by numerous scholars elsewhere.

Much academic engagement with cultural landscape over the last half-decade is framed as explicitly or implicitly "post-Sauer". Students and researchers tend to encounter his ideas as part of the concept's *history* rather than its present, overtaken by newer approaches. However, as we hope to show, the significant and robust critiques levelled at Sauer and his "morphology" should *not* cause us to wholesale reject his ideas. Nor should we underestimate the extent to which his perspective on landscape – an interactional, holistic context for human living – remains inherent to many works today. Therefore, before concluding, we engage something of Sauer's legacy and continuing promise for landscape studies.

The Limitations of Sauer's Morphology

There have been several critiques of Sauer's version of landscape. An early, and substantial, point of issue was the lack of definition around exactly what landscape *was*. Indeed, this endures as a weakness (or perhaps strength) of the term for many (see Jones, 2003). The geographer Richard Hartshorne, for example, in his 1939 survey text *The Nature of Geography*, argued that, whilst attractive for focusing attention towards material, visible scenes and constructing them, "in various undefined ways", as "a unitary concrete object", both landscape and the German term *landschaft,* upon which Sauer based his conceptualisation, were too vague to offer real analytical purchase (Hartshorne, 1939: 205). Landscape/*landschaft,* simultaneously evoking the perception (or representation) of land alongside its material features and content, was so ill-defined that there existed "almost as many gradations between [how such a term is used] as there are geographers who use it" (Hartshorne, 1939: 327). Hartshorne observed that "each geographer's definition of the word will vary according to what he thinks geography should study" (329). Sauer, in defining the term as he did, offered merely "the appearance of precision" without an actual basis for analysis (Hartshorne, 1939: 329).

Another critique by Hartshorne foreshadowed another debated deficiency of Sauer's morphology: his interest lay only in human activities as "material phenomena". Yet, much of what comprised social relations or culture, whilst no less "located", was "immaterial" (Hartshorne, 1939: 332). Thus, as later critics argued,

Sauer's approach to landscape was largely descriptive and positivist (i.e. based in the assumption of the accurate recording of observable, "true" facts). He was solely interested in material traces, promising little beyond "generalized description" (Mitchell, 2000: 103) and prioritising an accounting of the physical *record* of human–environmental interactions over and above how that environment was actively engaged with and lived in (see Hinchcliffe, 2003).

Consequently, Sauer did little to question how the context of relationships, beliefs, values, or practices of human groups served to *produce* landscape's material appearance. Indeed, Sauer's conceptualisation of "culture" in the cultural landscape and his inherited sense of culture as superorganic from the likes of Kroeber were especially attacked as researchers began to draw on work within areas, including cultural studies, Marxist political ecologies, and wider humanities writing influenced by post-structural thought. Partly, Sauer's focus on "archaic" rural landscapes was seen to risk ignoring the complex social, economic, and political forces presently at work in (re)making the cultural landscape. Moreover, the figure of the "human" that featured within his work, as an environmental actor, was painted as homogenous, passive, and lacking in historical or geographical difference (see Duncan, 1980; Mitchell, 2000: 35). His interest in the *material* character of landscape was as a succession of physical traces, rather than as a record of power relations, interests, or social struggle (see Chapter 3).

Ultimately, by framing culture as a force that operated separately and *above* the domain of material or lived relationships within landscapes, it was presented as a vaguely defined "way of life" without explanation nor examination in terms of its "inner workings" (Jackson, 1989: 19). Subsequent theorising of culture – incorporating attention to power relations, values, discourses, representations, ideologies, and even sensations or emotions – led to different accounts of landscape. By leaving culture alone, to scholars engaged with the new developments in cultural, social, and political theories of the latter-1970s and 1980s, Sauer's approach seemed theoretically underdeveloped, focusing too much on "the shape rather than the shaping of the earth" (Mitchell, 2000: 29).

It is worth remembering, as Mathewson (2018) notes, that such critiques often underestimated the changes in Sauer's work across his career. Tending to concentrate upon his "morphology" and early theorising of cultural landscape, there is less discussion of his more expansive treatments of land and life or how, particularly in his Latin-American-focused work, he examined the impacts of imperialism and colonialism upon societies and environments. Nevertheless, the above criticisms have inspired several of the cultural landscape approaches presented later in the textbook. For the remainder of *this* chapter, we consider how Sauer's ideas continue to influence studies of landscape *despite* their potential limitations.

The Legacy and Promise of the Sauerian Landscape

It can appear that Sauer's contribution to landscape is primarily historical: his work established the term as central to appreciating the historical relationship between human society and environmental change, but most of his specific ideas have been jettisoned following significant critique. Counter to this narrative, we emphasise three ways in which Sauer's approach remains important for studies of landscape today. Firstly, we outline the persistent influence of the "cultural landscape" concept within landscape policy and designations. Secondly, we suggest that a return to Sauerian studies of "land and life" in contemporary humanities and non-academic writing on place and environment is discernible. Third, we reflect on how Sauer's key proposals concerning landscape (an interactional entity, encountered phenomenally by the researcher as a "unit" of interest) remain detectable in contemporary social and cultural writing.

Landscape Policy

As Marc Antrop (2020: 6) notes, Sauer's influence has in part been that "landscape [...] was seen as a unique synthesis between the natural and the cultural characteristics of a region". The traces of his ideas can therefore be found in conceptualisations of landscape as both a defined region to be catalogued or classified and as a container for a particular "way of life". Such notions were subsequently combined within the Euro-American conservation movement (beginning in earnest in the latter-nineteenth century) as landscape became a central category for heritage and conservation policy and management during the twentieth century. Examples of such categories include the development of protected status designations, such as *Naturparks* and (since 2005) "National Nature Landscapes" in Germany and "Areas of Outstanding Natural Beauty" (AONB) (1956–2023) and "National Landscapes" (since 2023) in the UK.

Moreover, "landscape" (variously defined) has been incorporated as a key policy term by virtue of several multinational conventions (see Box 1.2). Such conventions and conceptualisations of landscape demonstrate both the enduring hold of the "cultural landscape" idea as well as the many ways in which such a term splinters along conceptual, policy, management, and applied lines. Indeed, amidst multiple frameworks, laws, designations, and policy positions, "the real landscape is often the only integrating concept" (Antrop, 2020: 12).

Box 1.2 International Landscape Definitions

Several international conventions enshrine definitions of landscape with real-world implications in terms of policy, management, and legal regulation. The following definitions provide examples and are taken from the United Nations Educational, Scientific and Cultural Organization (UNESCO) World Heritage Convention (1972) and the Council of Europe Landscape Convention (2000).

World Heritage Convention

> Cultural landscapes inscribed on the World Heritage List are cultural properties and represent the "combined works of nature and of man"

> [...] They are illustrative of the evolution of human society and settlement over time, under the influence of the physical constraints and/or opportunities presented by their natural environment and of successive social, economic and cultural forces, both external and internal. [...] The term "cultural landscape" embraces a diversity of manifestations of the interaction between humankind and the natural environment. [...] Cultural landscapes often reflect specific techniques of sustainable land use, considering the characteristics and limits of the natural environment they are established in, and may reflect a specific spiritual relationship to nature.
>
> (UNESCO, 2025: 74)

Council of Europe Landscape Convention

> "Landscape" means an area, as perceived by people, whose character is the result of the action and interaction of natural and/or human factors.
>
> (Council of Europe, 2000: 2)

Pause for thought: *Can you think of (or identify) any other definitions of landscape that shape the ways in which the environment is managed? Do they reflect, in any sense, Sauer's proposal of their "morphology" being the holistic outcome of human and nonhuman forces?*

The concept of landscape utilised in conservation and heritage is, evidently, more policy-oriented, socially descriptive, and applied in nature. It is, therefore, "less theoretical and academic" (Antrop, 2020: 12). Nevertheless, the definitions from which policy departs readily evoke a *morphological* sense of landscape as (variably) defined territory, perceived by an observing public, and made distinct by the interaction of "natural" and human worlds. Such definitions inform many of the legislative protections afforded to cultural environments today (such as Historic England's "historic landscape characterisation"); the landscape assessments that underpin them; the international networks (e.g. the Network for Landscape Conservation's "CIVILSCAPE" project) that have emerged to coordinate their management; and even the "landscape atlases" which typologise or document them. It is therefore important to consider contemporary landscape policy as indebted to morphological conceptions of cultural landscape.

Telling Stories of "Land and Life"

Over the past decade, Sauer's legacy has also been apparent across cultural geographers' engagements with creative or experimental forms of writing and telling. Dubbed the "telling turn" (see Lorimer and Parr, 2014), such work takes inspiration from various sources, not least an array of *literary* figures and texts (including British "New Nature Writing", discussed below), to showcase the value of a place-embedded, storytelling practice. One explicitly acknowledged (see Lorimer, 2013) point of departure for this kind of cultural geography scholarship is Sauer himself and the centrality of close-grained empiricism to his brand of landscape study. In work of this kind – often inductive, descriptive, and attentive to the character of the land and its etching by those who inhabit it – analysis and interpretation "does not always declare itself as such" and is often secondary, or "tucked away in the shadows of significant narrative detail" (MacDonald, 2014: 478).

Those engaged in this area are clearly influenced by a resurgence of regional and environmental nonfiction labelled "New Nature Writing" (NNR). This literary movement emerged in the latter-2000s, associated with writers including Robert Macfarlane, Helen Macdonald, Kathleen Jamie, Mark Cocker, and Richard Mabey. It also draws upon a longer lineage of popular environmental activism and consciousness emerging in the 1970s (see Smith, 2018) to express urgent, often elegiac attention to environmental subjects with an increasing awareness of landscape degradation as it attempts to promote ecological curiosity and reconnection (Cowley and Wylie, 2008). There often remains the sense that one cannot fully "know" its nonhuman objects of interest, demanding a humility on the part of the writer and an openness to different forms of knowing (Farrier, 2015). Writing attracting of the "NNR" label is often directed at a non-academic audience, aiming to enchant its readers, illuminate environmental histories, and democratise (or "counter-map") knowledges and experience (including the authors') of "nature" to encourage ethical engagement.

For geographers considering landscape, NNR has certainly inspired and often informs new efforts at "place-writing" or "geographical portraiture" (see Lorimer, 2013, 2019) that evoke the spirit of Sauer's morphology and emphasise the value of attending to "place as a subject in its own right" (Smith, 2018: 16). For example, Hayden Lorimer, in an essay on the presence of scarecrows in agricultural landscapes informed by both observations from train windows and active excursions into fields around his native Aberdeenshire, observes the manner in which this project owes a debt to Sauer:

> Where the divination of "geographical personality" was once suggested as the most rewarding outcome of regional fieldwork (Sauer 1941), in its stead might emerge "geographical portraiture" as a template for experimental landscape writing in the contemporary discipline. The list of scenic features as future subjects is near limitless.
>
> (Lorimer, 2013: 183)

In this way contemporary, often lyrical acts of storying the physical landscape; its histories and curiosities – the work of "bearing witness" to landscape amidst environmental upheaval (Daniels and Lorimer, 2012) – can be apprehended as another legacy of Sauer's espoused approach.

Ontological Remains

A third way in which Sauer's work endures, even despite his absence as an explicit point of reference or influence for contemporary scholars, is in the fact that landscape "enthrals" us as an ontological category. As a particular way of understanding reality, landscape serves to "hold things together" when researching or analysing cultural–environmental relations (Kirsch, 2015: 822). As Sauer's "unit concept", it designated a portion of space at once bounded by (a researcher's) interest and perception, defined by material characteristics produced via various interactions, and to be known *inductively,* rather than explained with reference to an imposed, external theoretical framework. Landscape as an ontological unit – a "thing" to be accounted for – remains often articulated in terms not *that* dissimilar from Sauer's. A "persistence of interest in landscape as an ensemble of objects, and not (only) as a picturesque visual scene or scenic representation" (Kirsch, 2015: 821) remains widespread.

For example, Vanessa Castán Broto (2019), in *Urban Energy Landscapes*, discusses the urban centres of Maputo (Mozambique), Concepción (Chile), Bangalore/Bengaluru (India) and Hong Kong (China). Drawing on field research that includes various qualitative, "talking" research methods as well as landscape observations gleaned through "walking transects", she examines the ways in which the relationships, debates, and experiences of energy politics are materialised in and through diverse physical urban landscapes and their constituent "energy artefacts". Across different locations, Castán Broto seeks not to predetermine nor stipulate the qualities of an "energy landscape". Instead, she considers how *specific* energy landscapes, comprising practices and artefacts associated with energy production, transmission, and use, exist as a "continuum of supportive structures that enable urban life", making up the "connective tissue" of urban activity (Castán Broto, 2019: 10–11).

In Castán Broto's book, then, landscape offers a lens for thinking about urban energy: its histories, politics, and social relations. It is "a spatial form that assembles things" (Kirsch, 2015: 822). "Landscape" takes human–nonhuman relationships and spatialises them (e.g. across the areal extent of an urban conglomeration) in terms of their productive interaction. The question of energy, and its historically contingent infrastructure or organisation, is rendered in terms of human relationships within a broader material environment that both shapes and bears traces of such activities. Indeed, the holism implied by "landscape", the fact that it asks us to consider things *together* rather than in isolation, is especially useful to researching the topic of energy. As Castán Broto writes, "a landscape perspective highlights the matrix that makes all those objects possible" as it does their accumulation through time and practice (2019: 49).

The framework of "landscape" therefore provides the research account with a scale of analysis that attends to "the environment in which life happens" (Castán Broto, 2019: 3). It is bounded by the researcher through their own phenomenal encounters with the world, its edges. For Castán Broto, acts of "following the energy" (e.g. on foot) involve "seeking different engagements with the city" and make "coherent" its various energy landscapes. In turn, energy landscapes emerge, inductively, through her research as "functional units" that "reveal a compendium of socio-energetic relationships that materialize within particular spatial arrangements in different choreographies of energy provision and use" (Castán Broto, 2019: 51). As she writes:

> In concrete places, such as the pathways alongside the main campus of the Universidade Eduardo Mondlane, a fence is erected as a linear object that embodies the administrative boundary. However, for the most part, neither administrative boundaries nor physical barriers across districts and energy landscapes are visible for the observer. Instead, a charcoal-based energy landscape is seamlessly transformed into a fully networked system – for example, along the long, congested avenues that cut across the city. The observer thus must perform an active role in boundary-making to define such landscapes.
>
> (2019: 50)

Specific energy landscapes are apprehended in terms of unique characteristics emerging as they are transformed towards the "purpose" of energy. The landscape is the *beginning* of this analysis, and the urban energy landscape is *the thing to be analysed*: an empirical object of interest that opens onto questions about energy politics, transition, or sustainability. An attentiveness to the "morphological" character of energy landscapes (their shape and history) is central. While this text mentions Sauer only twice and does not cite any of his writing, we can trace, in Castán Broto's work, the endurance of landscape as an ontological unit of analysis. Inductive, physical, and artefactual, it is to be examined through direct, field-based practice. Making such connections, we can understand part of Sauer's legacy more fulsomely: his defining of landscape as a fundamentally empirical, material, spatial, and phenomenal object for scrutiny.

Chapter Summary

Box 1.3 Key Concepts

Chapter 1

Environmental determinism
Landschaft
Landscape morphology
The superorganic
Interaction

This chapter has sought to consider the emergence of a "landscape perspective" for the study of human cultures and social relations within environments. In large part inspired by the ways in which contemporary scholars continue to articulate the value of a landscape approach as the premise, rather than end-point, of research, we have sought to illuminate both the origins of that perspective and its continued relevance, with a focus upon the work of Carl Sauer (Box 1.3). His morphological approach to the study of cultural landscape defined it as an investigation into how human societies transformed their surroundings, rather than being solely determined by them.

Although subject to robust critiques, a great many landscape scholars and professionals continue to think of landscape in ways not dissimilar to Sauer. Indeed, as Wylie (2007: 30) notes, "[s]uccessive waves of critique have had the paradoxical effect of establishing the Sauerian vision of cultural landscape as the legacy with which all others must grapple". Much of this critique is captured in the chapters to follow. Our argument here, however, is to acknowledge the idea of landscape that many people still deploy – that of an area. Sauer's ideas are akin to this more popular understanding of landscape and thus remain important, and perhaps always will be, because they are closer to the way that people (particularly those *not* linked to the academic study of cultural landscapes) think about this term.

Suggested Seminar Activities

Activity 1: Write down a definition of landscape as you understand it. Seal this into an envelope or put it in an email to yourself. When you have finished reading this textbook (or attending your university landscape course), open your definition and re-read it. How has your understanding of landscape *changed* as a result of what you have read or learned?

Activity 2: Identify an example of a landscape that contains:

a. evidence of both human and nonhuman processes shaping its features;
b. evidence of historical layers or changes in use/appearance over time;
c. the suggestion of having a particular, unique "character" as a consequence of (a) and (b).

In groups: share your examples with each other, and discuss the various human and nonhuman forces at work in shaping or producing these landscapes. Can you interpret any aspect of their form in relation to:

- The character of the environment influencing or prompting humans to engage with their surroundings in a particular way?
- Are humans adapting or changing the character of the environment to suit their own needs?
- The material traces of culture that human activities have deposited on the land.
- A notion of classifiable landscape "types" that might be common across the gathered examples, or that you are otherwise aware of?

Recommended Reading

Castán Broto, V. and Robin, E. 2023. Embracing change in infrastructure landscapes. *Landscape Research* 48(2): 165–173.

Lorimer, H. 2013. Scaring crows. *Geographical Review* 103(2): 177–189.

Mathewson, K. (2018). Carl Sauer and his critics. In: Denevan, W. and Mathewson, K. Eds. *Carl Sauer on Culture and Landscape*, Baton Rouge: Louisiana State University Press, pp. 9–28.

References

Antrop, M. 2020. A brief history of landscape research. In: Howard, P., Thompson, I., Waterton, E., and Atha, M. Eds. *The Routledge Companion to Landscape Studies.* Abingdon, London: Routledge, pp. 1–15.

Arntz, K. 1999. Landscape: A forgotten legacy. *Area* 31(3): 297300.

Castán Broto, V. 2019. *Urban Energy Landscapes*. Cambridge University Press.

Castán Broto, V. and Robin, E. 2023. Embracing change in infrastructure landscapes. *Landscape Research* 48(2): 165–173.

Council of Europe. 2000. *Council of Europe Landscape Convention.* Available at: https://www.coe.int/en/web/conventions/full-list?module=treaty-detail&treatynum=176, accessed on 20 August 2025.

Cowley, J. and Wylie, D. 2008. The new nature writing. *Granta* 102 (July 2008): 262–272.

Daniels, S. and Lorimer, H. 2012. Until the End of Days: Narrating Landscape and Environment. *cultural geographies* 19(1): 3–9.

Descola, P. 2013. *Beyond Nature and Culture.* Chicago & London: The University of Chicago Press.

Duncan, J. 1980. The Superorganic in American Cultural Geography. *Annals of the Association of American Geographers* 70(2): 181–198.

Farinelli, F. 2000. Friedrich Ratzel and the nature of (political) geography. *Political Geography* 19(8): 943–955.

Farrier, D. 2015. J.A. Baker's *The Peregrine* and its readers. *Interdisciplinary Studies in Literature and Environment* 22(4): 742–758.

Flowerdew, R. 2011. Torsten Hägerstrand. In: Hubbard, P. and Kitchin, R. Eds. *Key Thinkers on Space and Place.* 2nd Ed. London: Sage, pp. 199–124.

Gade, D. 2014. The continuing quest to understand Carl Sauer. *The AAG Review of Books* 2(3): 116–121.

Hartshorne, R. 1939. The nature of geography: A critical survey of current thought in the light of the past. *Annals of the Association of American Geographers* 29(3): 173–412.

Hinchcliffe, S. 2003. "Inhabiting" Llandscapes and Nnatures. In: Anderson, K., Domosh, M., Pile, S., and Thrift, N. Eds. Handbook of Cultural Geography. SAGE, London: Sage., pp. 207–226.

Hinchliffe, S. 2007. *Geographies of Nature: Societies, Environments, Ecologies*. London: SAGE.

Jacknis, I. 2002. The first Boasian: Alfred Kroeber and Franz Boas, 1896–1905. *American Anthropologist* 104(2): 520–532.

Jackson, P. 1989. *Maps of Meaning: An Introduction to Cultural Geography.* London: Unwin Hyman.

Jones, M. 2003. The Concept of Cultural Landscape: Discourses and Narratives. In: Palang, H. and Fry, G. Eds. *Landscape Interfaces: Cultural Heritage in Changing Landscapes*. Dordrecht: Springer, pp. 21–51.

Keighren, I. 2006. Bringing geography to the book: charting the reception of *Influences of geographic environment. Transactions of the Institute of British Geographers* 31(4): 525–540.

Kirsch, S. 2015. Cultural geography III: Objects of culture and humanity, or, re-'thinging' the Anthropocene landscape. *Progress in Human Geography* 39(6): 818–826.

Lorimer, H. 2013. Scaring crows. *Geographical Review* 103(2): 177–189.

Lorimer, H. 2019. Dear departed: Writing the lifeworlds of place. *Transactions of the Institute of British Geographers* 44(2): 331–345

Lorimer, H. and Parr, H. 2014. Excursions –Telling stories and journeys. *Cultural Geographies* 21(4): 543–547.

MacDonald, F. 2014. The ruins of Erskine Beveridge. *Transactions of the Institute of British Geographers* 39(4): 477–489.

Mathewson, K. (2018). Carl Sauer and his critics. In: Denevan, W. and Mathewson, K. Eds. *Carl Sauer on Culture and Landscape*. Baton Rouge: Louisiana State University Press, pp. 9–28.

Minca, C. 2007. Humboldt's compromise, or the forgotten geographies of landscape. *Progress in Human Geography* 31(2): 179–193.

Mitchell, D. 2000. *Cultural Geography: A Critical Introduction*. London: Sage.

Olwig, K.R. 1996. Recovering the substantive nature of landscape. *Annals of the Association of American Geographers* 86(4): 630–653.

Paasi, A. 2020. Regional geography. In: *International Encyclopedia of Human Geography*. 2nd Ed., vol. 11. Elsevier, pp. 309–320. www.sciencedirect.com/referencework/9780081022962/international-encyclopedia-of-human-geography

Qviström, M. and Wästfelt, A. 2020. In search of the landscape theory of Torsten Hägerstrand. *Landscape Research* 45(6): 683–686.

Ratzel, F. 2018. Lebensraum: A biogeographical study [1901]. *Journal of Historical Geography* 61: 59–80.

Sauer, C. 1925. The morphology of landscape. In: Leighly, J. Ed. 1969. *Land and Life: A Selection from the Writings of Carl Ortwin Sauer*. Berkley, CA: University of California Press, pp. 315–350.

Sauer, C. 1941. The personality of Mexico. *Geographical Review* 31(3): 353–364.

Semple, E. 1911. *Influences of Geographic Environment On the Basis of Ratzel's System of Anthropo-Geography*. London: Constable & Co.

Sharp, J. 2022. *Geographies of Postcolonialism*. 2nd Ed. London: SAGE.

Smith J. 2018. *The New Nature Writing: Rethinking the Literature of Place*. London: Bloomsbury.

Solot, M. 1986. Carl Sauer and cultural evolution. *Annals of the Association of American Geographers* 76(4): 508–520.

UNESCO. 2025. *Basic Texts of the 1972 World Heritage Convention, Edition July 2025*. Available at: https://whc.unesco.org/en/basictexts/, accessed 20 August 2025.

Vidal de la Blache, P. 15 July 1911. Les genres de vie dans la géographie humaine. *Annales de Géographie* 112: 289–304.

Winchester, H., Kong, L. and Dunn, K. 2013. *Landscapes: Ways of Imagining the World*. Harlow & London, UK: Pearson Education / Prentice Hall.

Wylie, J. 2007. *Landscape*. London: Routledge.

2 Landscape as a Way of Seeing

Introduction: What Do You See?

In 1994, Kentucky-based sociologists Thomas Grieder and Lorraine Garkovich tried to answer the question: why do different interpretations of landscape exist? Setting the scene for their argument, they offered the example of a hypothetical field, viewed by different individuals with different interests:

> Why does a real estate developer look across an open field and see comfortable suburban ranch homes nestled in quiet cul-de-sacs, while a farmer envisions endless rows of waving wheat and a hunter sees a five-point buck cautiously grazing in preparation for the coming winter? The open field is the same physical thing, but it carries multiple symbolic meanings that emanate from the values by which people define themselves. [...] These symbolic meanings and definitions are sociocultural phenomena, not physical phenomena, and they transform the open field into a symbolic landscape.
>
> (Greider and Garkovich, 1994: 1)

What Greider and Garkovich propose is that our surroundings are evaluated in ways reflecting the viewer's socio-cultural position, values, and relations. The environment (represented by the fictional field) is *seen* differently according to the priorities, interests, and values of the onlooker.

Consequently, many scholars have considered "landscape" as naming more than a unit of area or physical outcome of human–environmental relations (see Chapter 1). It also conjures the question of how we visually engage with the world. To use the word landscape evokes, immediately, a *scene* – often (not always) rural or "natural" – looked or *gazed* upon. Thus, cultural geographer Denis Cosgrove, whose influence is prominent within this chapter, defines landscape as "primarily geography as it is seen, imaged and imagined" (2003: 254).

And yet, as Grieder and Garkovich's example attests, there are many, competing and overlapping "ways of seeing". How, then, do these different looks of landscape emerge, dominate, or become taken for granted? After all, we encounter representations of landscapes everywhere, whether as oil paintings hung solemnly in art galleries; or the less celebrated, but no less conventional, images adorning

DOI: 10.4324/9781003172451-3

hotel lobbies and office work-spaces. Elsewhere, stylised or photographed environmental scenes proliferate in the pages of glossy magazines, advertisements, and computer desktops. What this chapter considers, then, is how "landscape" is constructed and encountered in dialogue with such representations, which reflect and reproduce the way we see and value our environments. In so doing, landscape is understood less as a *thing* in the world, and more as a *category of thought* by which we organise and make sense of space and existence.

Chapter Outline

The chapter begins with a discussion of vision and sight; specifically how vision can be construed as an active practice unfolding within specified social, cultural, and material contexts shaped by ideology. Developing this discussion of vision, the chapter moves to consider landscape as a visual social construction. In particular, drawing on the influential work of the "New Cultural Geography" scholarship produced during the 1980s and 1990s, an understanding of landscape as naming ways of seeing is outlined with reference to examples from landscape art in Europe and China. The influential definition of landscape in terms of iconography is examined before the final substantive section aims to qualify landscape as a visual category by foregrounding both important of feminist critiques and arguments for situating knowledge and the work of creative researchers who emphasise the embodied experience of seeing (and making art).

How Do We See the World?

Understanding landscape as a "way of seeing" the world means *de-naturalising* vision. Simply put, we need to stop thinking of seeing as something that *just happens* as a physiological, natural, objective activity. The adage "seeing is believing" positions vision as our unmediated relay between an external world and internal knowledge. And yet, as critical theorists have argued over many decades, seeing cannot be so easily separated from wider context.

From Vision to Visuality

Geographer Denis Cosgrove, influential in his discussion of the relationship between geographical knowledge and vision (e.g. Cosgrove, 2008), suggests one way to complicate the act of seeing: consider the language we use to describe doing it. Simply reflecting on the *way* we express the vision in language challenges a sense that sight "just happens" outside context or intention. As he writes:

> Language captures something of the rich cultural complexity of seeing. A glance is different from a stare, a sight is different from a vision. In considering the active use of the sense of sight most languages make a fundamental distinction between seeing and looking [...] The former suggests the passive and physical act of registering an external world by eye; the latter implies an intentional

> directing of the eyes. In English, viewing implies a more sustained and disinterested use of the sense of sight; while witnessing suggests that the experience of seeing is being recorded with the intention of its verification or subsequent communication. Gazing entails a sustained act of seeing in which emotion is stirred in some way, while staring holds a similar meaning by conveys a sense of query or judgement on the part of the starer. Such complexity in the ordinary language of seeing suggests something of its cultural significance in our relations with the external world.
>
> (Cosgrove, 2003: 253)

Moreover, one is hardly *born seeing* but evidently *learns to do so.* Consider, as Cosgrove observes, that babies are born without capacity to properly recognise shapes or objects, let along attribute social meaning to them. Light on the optic nerve is not *automatically* understood as "parent", "cup", "tree", etc. An infant comes to interpret patterns of light and colour, texture and surface as the brain develops. A child will often exclaim "what's that?", pointing towards something unfamiliar. Additionally, medical case studies of those who have lost their sight in early age and had it restored reveal how complex, traumatic, and bewildering the process of trying to work through the onslaught of visual stimuli can be (for example psychologist Oliver Sacks' [1995] account of patient "Virgil"). For those without *practice*, seeing is far from simple.

Neither is vision solely a matter of someone's *bodily* capacities. Many technological devices mediate how we see (as those wearing glasses or contact lenses know well). Importantly, though, telescopes, microscopes, cameras, mobile phones, televisions, and other objects change not only *how* or *what* we see; but also the very *meaning* of seeing. When eyes can be enhanced, or deceived, we think differently about the spaces, images, and environments we encounter. Consider how social media platforms like Instagram prompt a visual evaluation of urban landscapes as more or less "instagrammable" – in turn affecting not just how *we* interact with cities, but how cities are designed, managed, and marketed in the quest for "likes" and investment (Boy and Uitermark, 2017). When algorithms scan, classify, and circulate images based on patterns and likenesses, or the digital traces of previous preferences and judgements, we might consider too the impacts such developments have on how particular versions of "landscape" are represented, circulated, or reinforced (see Amoore, 2009; Bush, 2020).

Where do such considerations leave *vision*? In 1988, American art critic and historian Hal Foster suggested that instead of "vision" we should think in terms of ***visuality***. That is, consider not only the mechanics of seeing, but also "how we see, how we are able, allowed, or made to see, and how we see this seeing" (Foster, 1988: ix). Foster wanted to critically "socialise" and "historicise" seeing and its role in practices of power (for example via different "scopic regimes" or shared practices and assumptions of seeing and its meaning) as they informed a sense of self, identity, and society. His proposal has been taken up by many scholars, attending to vision as practical, learned, and disciplined. Indeed, as researchers, we might also reflect on how we are trained to see particular things in particular ways, in service to scholarly norms or assumptions (see Rose, 2003). These scholarly "visualities" are one amongst many different forms of seeing that jostle within any given context.

Social Construction and Representation

Recognising seeing as socially embedded follows from particular assumptions about reality and social existence. Not taking for granted that we simply see, but learn to do so in context, emphasises our world is not pre-given, *but* interpreted through shared "maps of meaning" (after Jackson, 1989), characteristic of specific places, periods, communities, social groups, politics, economies, belief systems, and lived experiences. This way of viewing the world is termed social constructivism and such an epistemology (i.e. our way of understanding knowledge or the means by which we know things) requires examining how things, actions, or relationships are imbued with meaning through the course of everyday life and interaction (see Berger and Luckman, 1967).

Thus, social construction frames a particular approach to the study of landscape, as Olaf Kühne (2019: 17) summarises:

> According to social constructivist landscape research, the construct "landscape" is the result of socially formed patterns of interpretation and evaluation, on the basis of which an internal synthesis of observed material objects and their connection with symbolic meanings takes place (for example, a stop sign is not merely a collection of metal and paint but communicates a general social expectation of action).

The aim of this kind of landscape study is to locate its understanding and associated activity within a wider cultural context and its relations of communication and power. Social constructionist approaches challenge the appearance of spaces or landscapes as "self-evident". There is, it is proposed, no "true" landscape; only historically contingent agreements concerning what "landscape" is understood to *mean*. Landscapes, as Greider and Garkovich (1994: 1) argue, are thus not simply "scenes" to view. They are "symbolic environments created by human acts of conferring meaning to nature and the environment, of giving the environment definition and form from a particular angle of vision and through a special filter of values and beliefs". Landscapes "are about *us,* rather than the natural environment": a reflection of cultural values, norms and beliefs and "the symbols and meanings that comprise [...] what people in cultural groups define to be proper and improper relationships among themselves and between themselves and the physical environment" (Grieder and Garkovich, 1994: 2). In this way, "landscape" slides from the *descriptive* to the *normative*. In other words: depictions, definitions, and discussions of what landscape "is" communicate not only *what is* but the way those things *should be* (or not).

Attention to how landscapes are depicted aligns with social constructionist interest in representations and their effects. Such representations include documents, texts, images, or other referential artefacts that somehow *stand in for* entities in the world. Simultaneously, representations complicate, mediate, and offer commentary or assessment of the value, significance, or meaning of their content (Casey, 2002). Thus, there is a *politics* to representation concerning who produces, circulates, and informs how representations are consumed, and under what conditions. This raises the question of representations as they relate to ideology: systems of ideas, beliefs, or values in society that propose particular interpretations or assumptions of the world and its working, often on behalf of or in the interests of those with power or authority. For the purposes of landscape, we might ask how powerful narratives manifest in portrayals of the environment that frame certain forms of land management, use, or ownership as positive, timeless, and proper.

Landscape as a "Way of Seeing"

One influential way that vision, culture, and ideology have been brought together and examined within studies of landscape has been via consideration of different "ways of seeing". This phrase, popularised by writer and critic John Berger (see Box 2.1), was applied to landscape by many in cultural-studies-influenced geography in the 1980s. For Berger, each image we encountered embodies a "way of seeing": it is a representation selected and composed according to cultural conventions and interests and interpreted in (specific) cultural and political context. Figures, including Cosgrove, Stephen Daniels, and Peter Jackson, took inspiration from this idea to examine how beliefs, values, and common assumptions characterising groups or societies were produced and circulated through acts of representation, in relation (or resistance) to dominant ideologies. Later work, incorporating post-structural theories concerning text and discourse (see Chapters 4 and 5) further elaborated how landscapes operated via acts of interpretation, practice, regulation, and resistance. Ultimately, the overarching claim that "landscape" fundamentally concerned questions of representation (e.g. of the environment) and its effects upon perceptions, evaluations of, and actions upon the world proved highly influential. It also served as a potent critique of the perceived overly-descriptive, apolitical approaches tied to the Berkley School (see Chapter 1). Below, we expand the way in which landscape was conceptualised as a "way of seeing", with a particular focus on how landscape art influenced perceptions of nature and the environment (e.g. see Daniels, 2004).

Box 2.1 John Berger's *Ways of Seeing*

In his book and TV series *Ways of Seeing*, critic John Berger (1926–2017) offered a critical, Marxist understanding of art in radical contrast to mainstream art history. He argued art needed to be understood not on its own terms, nor even those of the artist producing it, but in relation to a set of learned assumptions and historic, social conventions that informed how terms like "beauty", "truth", "genius", "civilization", or "taste" were applied and received. For Berger, such categories were neither innocent nor objective. They represented agreed values framing "ways of seeing" art (and wider culture) and serving "the ideological interests of the ruling class" (Berger, 1972: 86). Art history, Berger felt, had failed to consider art's *social meaning* within (and as productive of) wider historical context or ideology. As he argued: "The way we see things is affected by what we know or what we believe" (1972: 8). Thus, how we see, make sense of, and respond to the world reflects shared, cultural understandings of that world, its content, working, and meaning.

For Berger, technology was also crucial as modern techniques of copying, cropping, and circulation changed the meaning of art. With mass production and consumption, a painting of a specific landscape became "any place", as the portrait of a particular woman became a cipher for "beauty". More recently, visual artist Lewis Bush (2020) has engaged Berger's work to considering how our contemporary practices of seeing are further altered by algorithmic machine learning and big data in the context of internet search engines and social media, which, in their own way, "see" the world by scanning, comparing, and connecting images. As Berger argued, continuously encountering representations of the landscape adhering to set aesthetic conventions and being encouraged to evaluate subsequent representations on similar terms produces a shared visual grammar that affect how we comprehend the visual.

It is important to note that ways of seeing can *change*. Berger describes Medieval interpretations of fire imagery, once consumed in relation to ideas of hell and the devil, that might seem alien today. Thus, while "every image embodies a way of seeing, our perception or appreciation of an image [also] depends also upon our own way of seeing" (1972: 10). We, in turn, receive and interpret such images in dialogue with our own cultural conventions and experience. The art historian, Berger reasoned, should aim to interrogate the ideological function of vision.

Pause for thought: *Can you think of a contemporary technology that has been brought into existence since Berger set out his arguments in 1972 that has impacted how we see?*

Ideological Compositions

In 1984, Denis Cosgrove took inspiration from Berger, as well as the work of cultural studies scholars, to articulate a critical approach to landscape as a "way of seeing". His work was informed by figures such as Stuart Hall and Raymond Williams, adopting a Marxist stance towards theorising cultural values and practices. Williams, for example, examined the dichotomy of the rural and the urban as expressed within literature. He asserted the importance of an assumed *separation* between the "self-conscious observer" (often also a "self-conscious owner" who profited from others' labour rather than worked ground themselves) and the land as fundamental to the aesthetic characterisation of rural environments into "landscapes" (Williams, 1973: 172–178). Cosgrove, indebted to such work, proposed understanding landscape as a potent visual category reflecting (and emergent from) the socio-economic situation of landownership Williams described. In *Social Formation and Symbolic Landscape,* he argued "that the landscape idea represents a way of seeing – a way in which some Europeans have represented to themselves and others the world about them and their relationships with it" (Cosgrove, 1998: 1). Landscape was "not merely the world we see" but "a construction, a composition of that world" (1998: 13). Ideas *about* landscapes, their meaning and value, were entwined with both traditions in Western art and socio-economic and political change.

Connecting Art and Landscape Histories

Of particular interest for Cosgrove was the period extending from 1400 to 1900, during which societies in Europe had transitioned from a rural, feudal-agrarian model to an urbanising, capitalist-industrial one. As agricultural and industrial revolutions drove fundamental reorganisation of relationships between humans and the environment (in terms of administration and ownership as much as value), changes likewise occurred in how the environment was represented. Cosgrove traced the development of landscape painting from beginnings amidst the wealthy port-cities of Venice and Florence in the fifteenth century. There, affluent merchants commissioned artists to render realistic images of their property holdings, celebrating their status. Key to new ways of seeing and depicting landscape were also accompanying technologies and techniques of vision, such as the theory of perspective; new artistic media, optics, and lenses; and technologies of mass production. A similar tradition emerged in Flanders, Cosgrove argued, where, from the latter-fifteenth century, new methods of oil painting produced images better conveying the material heft and detail of lands (and objects) also being carefully mapped, catalogued, and measured as *properties* with saleable value.

Subsequently, Cosgrove documented new schools of landscape painting developing in the Netherlands during the seventeenth century, where great transformations were unfolding in the rural landscape. The concentration of landownership amongst a wealthy minority was naturalised by images that depicted the presence of working tenants on vast tracts as part of a timeless, natural order. At the same time in Italy, where the economic dominance of previous centuries had waned, artworks began to represent surroundings in a nostalgic light. French painter Claude Lorrain found inspiration beyond the urban centre of Rome where he lived. Drawing upon the iconography of an imagined, classical past, he conjured a lost, "golden age" of pastoral living in his works that focused upon scenes of poignant ruination or timeless, harmonious tranquillity all bathed in "mellow tones". Such scenes, peppered with motifs of the Grecco-Roman mythology, harkened to a rural ideal often referred to as "arcadia":

> The time is a golden age, an age of innocence and pure harmony between human life and the natural surroundings in which it is lived [...] the work is not intended to convey topographical accuracy. The landscape suggests above all a mood, a dream of classical innocence and warm perfection.
>
> (Cosgrove, 1998: 158)

These images were popular, offering "a vision of natural perfection" and an escapist fantasy. Tourists would seek attractive vistas so as to view them via tinted mirrors – "Claude glasses" (Figure 2.1) – that, with one's back to the scene in question, presented an image of the landscape in softer tones and colours evocative of Lorrain's original compositions (see Schama, 1995). Today, similar practices, via the use of photographic "filters" and effects, rework images in line with a set of aesthetic conventions (themselves informed by centuries of landscape art), produce depictions of landscape that meet wider social expectations and reproduce dominant cultural narratives (see Smith, 2021).

When, in the eighteenth and nineteenth centuries, English and French artists began to dominate the landscape genre, Cosgrove described how painters were reacted to the increasingly industrial, urban society of the time. Themes of beauty and nature's power marked a romanticising of wilder places as human influence over surroundings increased, and the growing numbers dwelling in cities were

Figure 2.1 A Claude Lorrain mirror in a sharkskin case. Wellcome Collection via Wikimedia Commons, CC BY 4.0.

further alienated and detached from the land (see Thomas, 1983). The notion of the "picturesque", influentially outlined by artist William Gilpin in 1792, proposed "natural" scenes be sought out and evaluated against artistic standards of beauty and composition. The art historian Malcom Andrews (1989), for example, links the emergence of "landscape tours" during the latter-eighteenth and early-nineteenth centuries around regions of Scotland, with scheduled stops and specially constructed viewing platforms to gaze onto notable scenes, to this way of seeing the environment.

Another influential notion, the "sublime", also circulated as it was articulated in writings by philosopher Edmund Burke and others in the mid-eighteenth century. The term moved from being a concept previously associated with religious experiences of awe and terror in the presence of God to a sense of being in the presence of awesome *nonhuman* natural forces or landscapes. Like the picturesque, searches for the sublime drove tourists to seek out and re-evaluate rugged, wild, apparently-untouched landscapes – previously maligned as unproductive or dangerous. Artists, in turn, supplied a stream of evocative imagery with individuals including Casper David Friedrich painting foreboding, crashing seas, cliffs, and mountains. The sublime infused an emerging mythologising of "wilderness" as the remnant of past, unspoiled nature to be cherished and conserved. Visiting the wild was figured as both religious and potentially therapeutic (see Chapter 12) in the face of accelerating modernity, urban squalor, and a sense of the closure (particularly in America) of those "frontier" landscapes where national identity (and masculinity) had been forged (Cronon, 1996).

In tracing these movements across of the history of landscape art, and situating them in the contexts of social and economic changes, Cosgrove proposed that "landscape" characterised a variety of "ways of seeing" the environment that should be examined "as part of a wider history of economy and society" and corresponding emergence of "elite consciousness". The ways in which "landscape" was depicted and understood propagated certain "political, social and moral assumptions" that became naturalised as aesthetic conventions (Cosgrove, 1998: 13). The history of this artistic genre, then, traced a societal response to the transformation of rural environments and emergence of a capitalist economic system that reshaped human relations to land and "nature", refigured as commodities (see Chapter 3). Importantly, landscape as a "way of seeing" needs to be appreciated as changeable and contingent, depending on cultural and historical context (see Box 2.2). It also could be unpicked and scrutinised in terms of a relationship between symbolism and iconography and wider worldly interests.

Box 2.2 Non-Western Ways of Seeing: Chinese Landscape Art

In his explorations of human practices of representing place, Edward Casey (2002) proposes that, in China, one can trace a historical development in landscape art mirroring some aspects of the Western tradition. A background "idealisation of place", as he characterises it, gives way to a foregrounding of environmental scenes as themselves the focus of representational acts during the Tang dynasty (618–906 CE). Of interest to Casey is the expression of this tradition in the eleventh century CE under the Northern Sung dynasty, and what it reveals of Chinese cultural understandings of nature.

In Western thought, Nature (capital "N") is an external "other" to the human – "something to take over [...] and then to take in" (2002: 95) by way of representation for the consumption of a detached observer. Conversely, artists, poets, writers, officials, and citizens in China conceptualised nature on different terms, as something "everywhere, equally so and at all times". The task of the landscape artist thus "is to convey this pregiven immanence": the ubiquitous happening of nature experienced as a reverberation, resonance, or "withness" in place. Such "withness" is captured, Casey describes, in the concept of "ch'i" – "the ever-circulating cosmic energy" conjoining and flowing through human and nonhuman.

Therefore, Northern Sung landscape painting attempts to enable the viewer to inhabit landscape and experience "an abiding unity with the natural world" (96). Certainly, this involves an established iconography of landscape's features: trees, rocks, as streams as "visual vocabulary" (2002: 96) helping the viewer recognise the depiction as a scene, rather than background "scenery". Indeed, for contemporary Northern Sung artists, such as Kuo Hsi, who Casey quotes at length, the painter's role was to "transmit" not a view but an *experience.* In this context, "landscape" understood as something in which the viewer was immersed; surrounding and sensed by the body (see also Chapter 8). The key objective of landscape *art* in this understanding was to communicate the "spirit" of – or what it was like to *be* in – place, rather than true likeness.

Pause for thought: *Have you ever encountered non-Western examples of landscape art or imagery? If so, how do these representations depict the landscape and what do they reveal about a relationship to "place" more widely?*

Consider, for example, the artwork produced by the collective of Aboriginal Australian Artists, Artists of Ampilatwatja, and their depictions of desert landscapes. How do such images convey place? See: https://ampilatwatja.com/

Iconography and Landscape

Cosgrove and others advocated studies of landscape via the scrutiny of representations. A collection of icons, symbols, and signs was perceived in ways reflecting specific "ways of seeing" that shaped perception of their coherence or supposed "unity" (see Kuhne, 2019: 49) as a "thing" to be given the label "landscape". Inspiration came from the toolkit of semiotics (see also Chapter 4) – and other approaches for deconstructing images, language, and meaning-making – to get beyond the *surface* of a scene and examine its construction to convey cultural narratives or values.

This approach is encapsulated within the influential collection *The Iconography of Landscape* (hereafter *TIOL*) (1988), edited by Cosgrove and fellow landscape scholar Stephen Daniels. Even decades later – and in the face of important criticisms considered below and in subsequent chapters – its arguments frame a key way of understanding landscape. The essays gathered within *TIOL* exemplify *critical* attention to the aesthetics of landscape as a matter of power and politics, and champion historically-attuned, interdisciplinary attention towards questions of environmental representation (della Dora et al, 2011).

In what may be the most cited sentence of their collection, Daniels and Cosgrove define "landscape" as "a *cultural image*, a pictorial way of representing, structuring or symbolising surroundings" (1988: 1 – emph. added). Crucially, this did not mean landscapes were immaterial. Indeed, representations were wrought "in a variety of materials and on many surfaces – in paint on canvas, in witing on paper, in earth, stone, water and vegetation on the ground". But whilst a landscape park could be "more palpable" than a "painting or a poem", it was "no more real, nor less imaginary" (1988: 1). The *meaning* of the former was bound up with the latter. Together, park and painting (or poem) (re-)produced and reinforced an idea of what a "landscape" was; further embedding meaning by "depositing yet another layer of cultural representation" (1988: 1). Like Berger's treatment of art, the iconographic approach of Cosgrove and Daniels understood landscape's meaning as historically and geographically specific. Thus, meaning was also fundamentally unstable given "our ability to invert signs and symbols, to recycle them in different context and thus transform their reference" (Daniels and Cosgrove, 1988: 7–8). The idea that some objective or empirical "truth" of landscape lay beneath its surface was rejected. Rather, they reasoned,

> landscape seems less like a palimpsest whose "real" or "authentic" meanings can somehow be recovered with the correct techniques, theories or ideologies, than a flickering text displayed on the word-processor screen whose meaning can be created, extended, altered, elaborated and finally obliterated by the merest touch of a button.
>
> (Daniels and Cosgrove, 1988: 8)

Dissecting Iconography and Vision in Practice

The iconographic approach involves careful consideration of how, to what ends, and with what effects representations of landscape are composed. Important work by Daniels, for instance, has examined the relationship between "ways of seeing" and the emergence of idealised park landscapes, with attention towards the organisation and laying out of lawns, vegetation, water features, and walkways revealing their accordance with aesthetic ideas and role in reproducing associated ideological assumptions in solid form (see Daniels, 1999; Daniels and Veale, 2015). Below, we consider an example from *TIOL*, concerning the construction of the Scottish

Highlands and its mythology (Pringle, 1988). An additional example, concerning ways of seeing mountain landscapes, is provided in Box 2.3.

In *TIOL*, Trevor Pringle (1988) offers a semiotic interpretation of the works of nineteenth-century landscape and portrait artist Sir Edwin Landseer; specifically, his paintings of highland landscapes commissioned by Queen Victoria. Deconstructing these landscape depictions, Pringle identifies their constituent "signs": the icons, motifs, or symbols that constitute key components of these images, carry, and communicating cultural associations or meanings. Together, as a depiction of landscape, these signs work together to imbue or associate the environment with particular ideas, values, or properties (see Chapter 4). In this way, Pringle identifies in Landseer the roots of a "highland myth": a powerful set of associations attached to the environment of Scotland that continues to shape how it is managed today (see MacDonald, 2023). The artist's portrayals of unpeopled vistas, rolling hills, and tranquil glens, amidst which the royal family abide as though part of an idyllic "natural order", serves to remove Scottish nature from the historical, economic, and political context producing of it (Pringle, 1988: 143).

Examining one of Landseer's most famous images, *Royal Sports on Hill and Loch* (1874), depicting the Queen arriving by boat and greeting her husband Prince Albert in highland dress, Pringle describes how Victoria intended the image to capture "a beautiful exemplification of peaceful times, and of the independent life we lead in the dear highlands" (Pringle, 1988: 114). The image reveals the Queen's desire to celebrate and solidify her affiliation with the Highlands (having first leased Balmoral Castle in 1847) as an empty, pristine wilderness to be enjoyed by and for royal hunting pursuits. In turn, the image denies the 1715 and 1745 Jacobite rebellions; the subsequent dismantling and banning of Scottish clan culture (until 1780); and the violent clearances removing farming tenants from the land to make it appear thus. Likewise, another of Landseer's images, *The Highland Lassie* (1851), intended to be "all peace and sunshine", depicts the Highlands as rural arcadia where "sheep graze peacefully in the background while, in the distance, smoke spirals upwards into a serene sky from [...] happy homes" (Pringle, 1988: 149). Again, such imagery exorcises lived history and awkwardly contrasted wider industrial urbanisation that had reduced "the great majority of Scots into landless proletarians" (Pringle, 1988: 153). Less wild and Edenic, Scotland in the mid-nineteenth century was characterised by "profound social and economic crisis", the "harsh realities" of which were "veiled" by a "Highland myth" which "helped propagate an increasingly false regional image of a tranquil rural Scotland" (1988: 153).

Nevetheless, Landseer's images were sought after by a middle-class urban market, keen to consume representations of wild, romantic nature. Indeed, such a "way of seeing" Scotland, and associated iconography, endures in the symbolism of today's tourism promotion (as well as the politics of landownership and management). As Fraser Macdonald succinctly puts it: "our ecological problems are in some sense art history problems, caught up with what we think Scotland used to look and feel like, and what it should look and feel like" (Macdonald, 16 November 2023). Work like Pringle's remains relevant in illustrating how particular ways of seeing landscape and nature serve the ideology of those in power. After all, "[t]he question of who owns the land is connected to questions of the way places are characterised, and whose interests these characterisations serve" (Macdonald, 16 November 2023). Calls to reform land law in Scotland, where around 433 individuals and companies own half of the country, have struggled to make headway for some time (see Carrell, 23 March 2024). In romanticising the Highland landscape, "certain classes of people have signified themselves and their world through their imagined relationship with nature [...] through which they have underlined and communicated their own social role and that of others" (Cosgrove, 1998: 15).

Box 2.3 Visualising Mountains through History

Tracing a history of mountains as visual objects, Veronica della Dora describes their transformation from symbolic icons of memory and religious experience to vantages from which one achieves mastery over a scene to be viewed. As she describes, throughout Western history, "[m]ountain encounters have helped define particular ways of seeing, experiencing and representing the world" (della Dora, 2017: 107). This is in part because mountains are key constituents of the viewed landscape. Their transformation from jagged, threatening, barren peaks infested with bandits and devils; to romantic summits enabling encounters with sublime nature traces the history of wilderness myth. Simultaneously, they offer a viewpoint *onto* the landscape; the vista from the mountain becomes "an approximation of divine knowledge" (della Dora, 2018: 190).

Medieval artists' depictions of religious scenes, such as Jesus's fabled encounter with the Devil in the wilderness, feature mountains represented as subordinate (and small) compared to the sacred or wicked figures towering over them. In such works, della Dora argues that the mountain landscape serves as "a memory theatre [...] a container of symbols and allegories" (2018: 193–194), rather than the focus of the scene. However, when Renaissance techniques of linear perspective proposed a view "mastered by a mono-focal gaze from a fixed point of view" (2018: 198), such ways of seeing mountains were displaced. Mountain landscape scenes featured in the background of famous works by Leonardo Da Vinci (e.g. the Mona Lisa, 1503–1514) to give a sense of scenic depth. Indeed, the measurement of mountains and their capacity to host an authoritative "birds eye view" informed such ways of seeing the landscape and later "cosmographic" paintings presented a "god's eye" view from above over creation as the means to bask in its sublimity.

Considering the ascent of Mont Blanc by Horace Benedict de Saussure in 1767, della Dora notes the reframing of the mountain landscape via the concept of the "panorama" as these environments were reconstructed in the cultural imagination as sites for scientific inquiry. Saussure's "fisheye" views of the Alps, sketched from a number of summits, capture this scientific ambition of "unity and panoptic control" (2018: 205): a world organised around a central, authoritative observer. The rotunda (a circular building depicting a 360° view of a scene) would become a popular installation in many towns, offering "the illusion of total immersion" (della Dora, 2018: 206) and framing panoramic ways of seeing for a public without requiring ascents to the mountain top. In this way, mountains moved into cities, as city dwellers travelled to mountains. The view from the summit "transformed the world into an exhibition [...] centred on a spectator paradoxically immersed and at the same time detached from it" (della Dora, 2018: 207).

***Pause for thought**: Can you think of another landscape type or "object" that is often represented in the same kind of way across social media today? Has this particular kind of landscape or landscape object always been represented in this way?*

Situating the View(er)

This chapter has considered at length the conceptualisation of landscape as "way of seeing": a visual encounter with an environment reflecting context and shared cultural values. Attention in landscape studies towards the construction of environments via representations and discourse (see Chapters 4 and 5) highlights how particular cultural conventions, values, beliefs, and norms come to be *naturalised* – taken for granted – as the basis for making sense of the world. However, a key critique of such approaches concerns the need to *situate the viewer*, as well as what is viewed (and how), in the social world. This reflects broad-scale acknowledgement across the social sciences and humanities that academic knowledge-making is neither inseparable, nor immune, from the circumstances of its production.

Feminist writers have been especially influential in asserting the partiality, and situatedness of such knowledge (Haraway, 1991; Rose, 1993) to emphasise that the representations, texts, or images of landscapes we encounter did not document views, attitudes, or perspectives held (or similarly received) by all. Consequently, increased awareness of the ways in which those who are not male, white, or English-speaking have been historically (and continually) excluded from knowledge-making challenges any presumption that the way in which we perceive or experience the world is universal – some often not considered the work discussed so far in this chapter. It is important that representations of the world are considered not only in terms of their effects, but also in terms of *who* authored them, and *whose view* they depict (or omit).

Who Sees Landscape?

In *Feminism and Geography*, Gillian Rose considers landscape as "a central term [...referring] to one of [geography's] most enduring interests": the relationship "between Nature and Culture" (1993: 151). And yet, for Rose, knowledge (especially geographical knowledge) of landscapes was inherently partial. She, and other scholars of landscape, would draw on the work of feminist philosophers, particularly Donna Haraway (1991). In her writing on animal science, particularly primatology, Haraway emphasised that, rather than reflecting what philosopher Thomas Nagel termed the "view from nowhere", scientific knowledge was a narrative practice of constructing accounts of and about the world. Thus, she argued, knowledge (whether scientific or vernacular) was always inherently *bound up with* the context of its creation. It was *situated* rather than detached from place since knowledge claims were always the work of particular people, in particular surroundings, producing an account of their world. Removing knowledge from the context of its production, Haraway argued, performed a "god trick of seeing everything from nowhere" and erased our own *involvement* in the production of accounts about the world (1991: 189). In turn, we forget that certain claims, widely accepted as universal, objective, and true, are often those of a very particular group of individuals. For Rose, the prominence of white male academics working to define and explain cultural landscape had meant the role of *gender* had been too little considered, whilst a male way of seeing and evaluating landscape was treated unproblematically.

For example, Rose highlights that many aesthetic frameworks regarding the representation of landscape in the eighteenth and nineteenth centuries – romanticism, the pastoral etc. – were often also associated with a particular gender politics. As she writes: "[i]deas about natural order were epitomized in the 'natural' difference between men and women" and a naturalising of the latter's social role (1993: 165). Similarly, both environments and women's bodies were depicted as "enticing and inviting [...] to be explored, mapped, penetrated and known" (Rose 1993: 164); as well as framed via notions of beauty, productivity (or fertility), or the maternal (i.e. "mother" nature). Thus, such images mapped a "sensual topography of land and skin" (Rose, 1993: 169).

To elaborate this point, Rose revisits an oft-discussed image, Thomas Gainsborough's *Mr and Mrs Andrews* (c. 1750 – see Figure 2.2). Considered to exemplify the relationship between ideas of landscape and property ownership, it portrays the titular wealthy, aristocratic couple at home amidst a scene of pastoral beauty (their landholdings) (see discussion by Berger, 1972; and Daniels and Cosgrove, 1988). Rose offered a complication of that reading, however, that hinged upon prizing the couple *apart* and attending to how their gender afforded them different relationships with the surroundings in which they sat. As she writes of the work:

> His potential for activity, his free movement over his property, is in stark contrast not only to the harsh penalties awaiting poachers daring the same freedom of movement over his land [...] but also to the frozen stillness of Mrs Andrews. Moreover, the shadow of the oak tree over her refers to the family tree she was

Figure 2.2 Mr and Mrs Andrews (c. 1750) by Thomas Gainsborough. National Gallery via Wikimedia Commons, public domain.

> expected to propagate and nurture; like the fields she sits beside, her role was to reproduce, and this role is itself naturalized by the references to trees and fields […] Mr Andrews is represented as the owner of the land, while Mrs Andrews is painted almost as a part of that still and exquisite landscape.
>
> (Rose 1993: 160–161)

In the same period as this painting was produced, arguments were being consolidated in Western scientific discourse that framed women as more "natural beings"; less rational and less intelligent. Men, meanwhile, were feted as possessors of true reason and rationality. Thus, to *situate* this image in the context of its time recognises that its depiction of landscape works to encode these same gender relations. After all, it was not the case that they were *both* landowners: according to the laws of the mid-eighteenth century, only Robert Andrews *could* own land. Indeed, much of the estate pictured had come to him in the form of a dowry from the family of his wife, Frances Mary Carter.

Ultimately, understanding such context is vital. It reveals the many alternative perspectives on landscape occluded by the white, male gaze of much landscape study. Rose's contribution remains influential in checking our tendency to "retreat to a critical distance" (Rose, 1993: 101) when analysing landscape representations – indeed, visual materials more broadly (see Rose, 2022). Rather, we must acknowledge the situatedness and positionality of those viewing (and writing about) landscape; foregrounding the (material, bodily, cultural) circumstances under which claims about the world are produced. Attention to these concerns allows us to "imagine different kinds of landscape" (Rose, 1993: 156) beyond those traditionally defined (further explored in Box 2.4). It also allows us to examine the active role of situated *viewers* in making sense of the scenes before them. For example, scholars considering audience responses to visual artistic works such as land art and sculpture have demonstrated "the active role of the audience in an art encounter" (Warren, 2013: 94). In turn, social or cultural background and context are differently brought into play within these encounters by pieces serving as "a prism where art, landscape, sky, and its human and non-human inhabitants act upon each other" (Warren, 2013: 96). These kinds of arguments situate our ways of seeing within ongoing relations that involve our bodies and their social relations in different ways (see Chapter 8).

Box 2.4 Alternative Landscape Visions

Geographer Divya Tolia-Kelly's work with British Asian women in the UK (see also Chapter 7) reveals the ways in which mobility and racialised experience in the context of Empire feeds into postcolonial landscape sensibility as a composite of lived and cultural memories of "home", variously located. In her research, she has used visual arts-based practice to capture this community's distinctive "ways of seeing" British landscapes conventionally aestheticised as wild and romantic (such as the Lake District). In the gaze of these women, such scenes are rendered less familiar, or beautiful, becoming boring, barren, or fearful (see Tolia-Kelly, 2007).

An additional example that sees questions of racial experience and identity complicate and challenge conventional aesthetic ideals of landscape is evident in the work of American portrait artist Kehinde Wiley. Wiley has often worked to disrupt the traditional aesthetics of "old masters" by referencing and reworking classic European paintings and scenes such that their original subjects are replaced by contemporary Black individuals from the present day. His 2021 exhibition, *The Prelude,* at the National Gallery saw him apply such an approach of "part quotation, part invention" (see Kehinde Wylie, n.d.) to iconic works of land- and seascape art by Caspar David Friedrich and others. Accompanying Wylie's images was a video installation depicting Black residents of London met by the artist near the gallery itself and taken to Norway to explore and engage with its snowy, mountainous environment. Together, these works put contemporary figures of "Blackness" into conversation with traditionally white, European, colonial visions of the landscape to create "a provocation, a look at what is possible" that also considers how representations of the environment centre, celebrate, or exclude.

- *See also:* https://kehindewiley.com/exhibitions/the-prelude/

Pause for thought: *What do you think of Kehinde Wylie's images? What do you think the effect of these images is upon the way we picture ourselves in relation to landscape?*

The Practice of Painting

Further to calls to situate the work of representation within social contexts, contemporary scholarship on landscape has also sought to consider art as a practical, *embodied* activity that takes place in the world. In this sense, representing a scene in paint or ink becomes less a matter of faithfully documenting what is "seen" than "a situated means to know" (Hawkins, 2015: 263): one practice amongst many by which we try to make sense of the landscape.

Creative work within cultural geographies has championed the use and promise of arts-based methodologies for researching human–environment encounters (see Hawkins, 2014). This prompts consideration of the embodied nature – or phenomenology – of artistic activities (see Chapter 8) of art "as a practice" that we, whether maker or viewer, "actively and physically 'do'" (Miller, 2017: 246). Attention to art's "doing" reveals *how* landscapes are visually constructed through practical engagement. One need not be skilled to engage such research methods as the act of *trying*, and moments of "ineptitude", can generate valuable insights into how the world is, actively, seen (Hawkins, 2015).

As an example, the recent work of geographer John Wylie, in collaboration with Swansea-based artist Catrina Webster, considers experience of drawing and painting as a prompt towards a more bodily and felt account of vision. Sketching and painting the Hoopern Valley in Southwest England, the UK, the experience of using pen, paint, easel, and paper, and the way these objects together, in hands and on pages, mediate and transform the encounter with landscape, is explored. As the authors write, "the canvas was as active and as ever-present an element as the ball is in a football match" (Wylie and Webster, 2019: 37). Indeed, through sketching, the fluidity of the scene and unreliability of the eye become evident; the landscape seems to "quiver" in the shuttling of gaze between world and canvas and refuses to be held still. Consequently, drawing and seeing becomes a physical and emotional activity. Learning to paint or draw requires one "to learn *a new way of seeing,* and in doing so to be drawn out, drawn in – drawn *closer* to visible objects, patterns and relations" (Wylie and Webster, 2019: 38). Considering such practices in relation to other media – for example ceramics (Miller, 2017) or sculpture (Warren, 2013) – further complicates the business of seeing and emphasises the role of the body in producing (or performing) aesthetic conventions and ways of seeing.

Chapter Summary

> **Box 2.5 Key Concepts**
>
> **Chapter 2**
>
> *Social Construction*
> *Ways of Seeing*
> *Ideology*
> *Iconography*
> *Situated Knowledge*

This chapter has considered the study of landscape as a specifically *visual* cultural category. Understood this way, landscape constitutes a "way of seeing" (or evaluating) the environment with reference to aesthetic conventions that are heavily indebted to context and history (including of art and aesthetics); as well as political, economic, and social relations and interests. Some of the key concepts that underpin this approach to the study of landscape are highlighted in Box 2.5.

Why does this approach matter? As discussed regarding the Highland Myth, historically-defined ways of seeing landscape continue to influence us today. Consider the ways in which places like Yosemite and other American National Parks are managed. Often this is in an ecological state reflecting less how they may have been maintained by Indigenous communities, such as the Ahwahnechee, prior to European colonisation, and more with reference to *aesthetic* notions of wilderness, proffered in nineteenth-century accounts by figures such as John Muir or paintings by artists like Albert Bierstadt (see Ellison, 2013). Thus, historic representations of landscape frame ways of seeing, which in turn shape how we act upon (or manage) the environment today.

Importantly, the chapter also considered the need to recognise that ways of seeing landscape are not universal but *situated.* We must ask how we see, but also *who* sees landscape. We can consider whose ways of seeing dominate and why; and how *alternative visions* can be uncovered or promoted, revealing different ways of valuing or engage with surroundings. Situating the act of vision also invites examination of how seeing (and representing) landscape involves the immersion of bodies in the material environment (see Chapter 8). Thus, the application of creative methodologies to landscape studies can elaborate the relationship between depicting and experiencing the world.

Suggested Seminar Activities

Activity 1: Using whatever device (smartphone, digital camera etc.) and applying whatever filters, enhancements, or edits or you wish produce a photo of a "landscape" and bring it to class. You may use an image posted to a social media platform (e.g. Instagram) or borrow a camera/smartphone to take the image, so long as you can bring a copy of the image to class.

Be prepared to discuss in groups:

- What landscape is depicted? How have you framed the scene?
- Why did you select the landscape you did?
- What (if anything) did you do to enhance the image afterwards, and why?

Compare the images taken by those in your group:

- How many of the images:
  - Are oriented in "landscape" format? If there are any that are framed differently, why is this?
  - Feature people only as distant or side-lined figures, if at all, rather than as the focus of the image?
  - Contain: more nonhuman nature than human-built structures, rural scenery, big skies, sunsets, watercourses, mountains, forests or oceans, notable lines or linear objects that guide the gaze towards a distant vanishing point?
- Are there other common features or noticeable differences between the images?
- Reflecting on the activity as a group/class: What do the images taken reveal about how you (individually, collectively) define landscape in visual terms? What do they reveal about *your ways of seeing?*

Activity 2: Using any social media platform where images are routinely shared and associated with hashtags: search for images associated with the hashtag of your country (e.g. "#greatbritain") and the hashtag "#landscape".

Consider the images that appear:

- What environments are being depicted? What environments are not being represented?
- What significance (for national identity, politics, history, environment), if any, is attached to the landscapes being depicted?
- How many images have been shared for evidently economic or commercial purposes (e.g. business promotion and tourism)?
- What kinds of judgements, values, or associations (negative or positive) about your country are being produced or reinforced through these images?

Recommended Reading

Cosgrove, D. 2003. Landscape and the European sense of sight – Eyeing nature. In: Anderson, K., Domosh, M., Pile, S. and Thrift, N. Eds. *Handbook of Cultural Geography.* London: Sage, pp. 249–268.

della Dora, V. 2018. Mountains as a way of seeing: From mount of temptation to Mont Blanc. In: Kakalis, C. and Goetsch, E. Eds. *Mountains, Mobilities and Movement.* London: Springer, pp.189–211.

Smith, S. 2021. Landscapes for “likes”: Capitalizing on travel with Instagram. *Social Semiotics* 31(4): 604–624.

References

Amoore, L. 2009. Lines of sight: On the visualisation of unknown futures. *Citizenship Studies* 13(1): 17–30.

Andrews, M. 1989. *The Search for the Picturesque: Landscape Aesthetics and Tourism in Britain, 1760–1800.* Stanford, CA: Stanford University Press.

Berger, J. 1972. *Ways of Seeing.* London: Penguin.

Berger, P and Luckman, T. 1967. *The Social Construction of Reality: A Treatise in the Sociology of Knowledge.* London: Penguin.

Boy, J. and Uitermark, J. 2017. Reassembling the city through Instagram. *Transactions of the Institute of British Geographers* 42(4): 612–624.

Bush, L. 2020. *Ways of Seeing Algorithmically: John Berger Reloaded.* Paris: Editions Trocadero.

Carrell, S. 23 March 2024. Land ownership in rural Scotland more concentrated despite reforms, study finds. *The Guardian*, online. Available at: www.theguardian.com/uk-news/2024/mar/23/land-ownership-in-rural-scotland-more-concentrated-despite-reforms-study-finds. Accessed on 6 June 2024.

Casey, E. 2002. *Representing Place: Landscape Painting & Maps.* Minneapolis, London: Minnesota University Press.

Cosgrove, D. 1998. *Social Formation and Symbolic Landscape.* London: University of Wisconsin Press.

Cosgrove, D. 2003. Landscape and the European sense of sight – Eyeing nature. In: Anderson, K., Domosh, M., Pile, S. and Thrift, N. Eds. *Handbook of Cultural Geography.* London: Sage, pp. 249–268.

Cosgrove, D. 2008. *Geography & Vision: Seeing, Imagining and Representing the World.* London: I.B. Taurus.

Cronon, W. 1996. The trouble with wilderness or, getting back to the wrong nature. *Environmental History* 1(1): 7–28.

Daniels, S. 1999. *Humphry Repton: Landscape Gardening and the Geography of Georgian England.* London: Yale University Press.

Daniels, S. 2004. Landscape and art. In: Duncan, J., Johnson, N. and Schein, R. Eds. *A Companion to Cultural Geography.* Oxford: Blackwell, pp. 430–446.

Daniels, S. and Cosgrove, D. 1988. Introduction: Iconography and landscape. In: Cosgrove D and Daniels S Eds. *The Iconography of Landscape.* Cambridge: Cambridge University Press, pp. 1–10.

Daniels, S and Veale, L. 2015. Revealing repton: Bringing landscape to life at Sheringham Park. *Landscape Research* 40(1): 5–22.

della Dora, V. 2017. *Mountain.* London: Reaktion.

della Dora, V. 2018. Mountains as a way of seeing: From Mount of Temptation to Mont Blanc. In: Kakalis, C. and Goetsch, E. Eds. *Mountains, Mobilities and Movement.* London: Springer, pp. 189–211.

della Dora, V., Lorimer, H. and Daniels, S. 2011. Classics in human geography revisited: Cosgrove, D. and Daniels, S. Eds. (1988) *The Iconography of Landscape: Essays on the Symbolic Representation, Design and Use of Past Environments*. Cambridge: Cambridge University Press. *Progress in Human Geography* 35(2): 264–270.

Ellison, A. 2013. The suffocating embrace of landscape and the picturesque conditioning of ecology. *Landscape Journal* 32(1): 79–94.

Foster, H. Ed. 1988. *Vision and Visuality*. Seattle: Bay Press.

Greider, T. and Garkovich, L. 1994. Landscapes: The social construction of nature and the environment. *Rural Sociology* 59(1): 1–24.

Haraway, D. 1991. *Simians, Cyborgs and Women: The Reinvention of Nature*. New York: Routledge.

Hawkins, H. 2014. *For Creative Geographies: Geography, Visual Arts and the Making of Worlds*. Abingdon: Routledge.

Hawkins, H. 2015. Creative geographic methods: Knowing, representing, intervening. On composing place and page. *Cultural Geographies* 22(2): 247–268.

Jackson, P. 1989. *Maps of Meaning: An Introduction to Cultural Geography*. London: Routledge.

Kehinde Wiley: The Prelude. n.d. News, *Stephen Friedman Gallery* website. Available at: www.stephenfriedman.com/news/476-kehinde-wiley-the-prelude/. Accessed on 6 June 2024.

Kühne, O. 2019. *Landscape Theories: A Brief Introduction*. New York: Springer.

MacDonald, F. 16 November 2023. Balmorality. *London Review of Books* 45(22). Available at: www.lrb.co.uk/the-paper/v45/n22/fraser-macdonald/diary. Accessed on 6 June 2024.

Miller, A. 2017. Creative geographies of ceramic artists: Knowledge and experiences of landscape, practices of art and skill. *Social & Cultural Geography* 18(2): 245–267.

Pringle, T. 1988. The privation of history: Landseer, Victoria and the Highland myth. In: Cosgrove, D. and Daniels, S. Eds. *The Iconography of Landscape*. Cambridge: Cambridge University Press, pp. 142–161.

Rose, G. 1993. *Feminism and Geography*. Cambridge: Polity Press.

Rose, G. 2003. On the need to ask how, exactly, is geography "Visual"? *Antipode* 35(2): 212–221.

Rose, G. 2022. *Visual Methodologies*. 5th Ed. London: Sage.

Sacks, O. 1995. *An Anthropologist on Mars: Seven Paradoxical Tales*. New York: Alfred A. Knopf.

Schama, S. 1995. *Landscape and Memory*. London: Harper.

Smith, S. 2021. Landscapes for "likes": Capitalizing on travel with Instagram. *Social Semiotics* 31(4): 604–624.

Thomas, K. 1983. *Man and the Natural World: Changing Attitudes in England 1500–1800*. London: Allen Lane.

Tolia-Kelly, D. 2007. Fear in paradise: The affective registers of the English lake district revisited. *The Senses and Society* 2(3): 329–351.

Warren, S. 2013. Audiencing James Turrell's *Skyspace:* Encounters between art and audience at Yorkshire Sculpture Park. *Cultural Geographies* 20(1): 83–102.

Williams, R. 1973 [2016]. *The Country and the City*. London: Vintage.

Wylie, J. and Webster, C. 2019. Eye-opener: Drawing landscape near and far. *Transactions of the Institute of British Geographers* 44(1): 32–47.

3 Landscape and Labour

Introduction: Landscape as Labour

Outside of Edinburgh, Scotland, lie some fairly unique landforms (Figure 3.1). Red in hue, with steep sides and flat tops, these protuberances rise out of the undulating scenery and sit, unusual but unnoticed, amidst the landscape of the expanding city. They are the so-called "bings": spoil heaps from the once burgeoning, but hugely wasteful, shale oil industry; dominating this area from the 1860s until its gradual decline almost a century later. The spent shale is a physical reminder of the relationship between this landscape and labour: refineries now silent, jobs now lost, and a physical environment transformed.

The bings exemplify how landscapes are produced, actively, through work. This work does not stop with their creation – they remain, affecting how that landscape might be worked on, or valued, later – and they are constantly subject to various forms of ongoing action. For a long time, these landscape features were disliked as ugly reminders of past industry. As Cal Flyn observes, "no one wants to be defined by their spoil heaps" (2021: 17). Consequently, some were levelled. Others were mined for "blaes" (a reddish substance excavated from the bings for use as construction material) to make building blocks, motorway landfill, and even surfaces for all-weather sports pitches. There were once 27 such bings, 19 of which remain.

Other interests, however, have sought to preserve them for their historic value as examples of industrial and cultural landscape heritage. One effort to do so involved reframing the bings as unintentional landscape sculptures, collectively entitled *Niddrie Woman.* This recontextualisation of the spoil heaps as heritage came before post-industrial landscapes were offered legal protection (see Richardson, 2012). The mounds have also been proposed as "island refugia" (Harvie, 2004): inadvertent ecological niches providing a haven for biodiversity; a side effect of their societal neglect. Moreover, the later creation of a 16 km "Shale Trail", combining a walking and cycling route and culminating in a museum documenting the labour history of the bings, was launched as part of the most recent effort to articulate their value, which "cannot easily be measured"

DOI: 10.4324/9781003172451-4

Figure 3.1 Five Sisters Shale Bings, West Lothian, Scotland, UK. Image by *tormentor4555*. Wikimedia Commons, CC BY-SA 3.0.

(shaletrail.co.uk, 2021). So, labour in the past produced these mounds, which continue to exert a force over the material form and meaning of landscape in the present (and future) as their value shifts and is struggled over; or new proposals for their removal or preservation come to the fore.

This understanding of landscape as in a historical process of formation bound up with labour, and the reasons why such landscapes change or endure, form the focus of this chapter. We thus move from an interest in landscape constructed through acts of visual appraisal (whether by scholar or artist) towards an examination of landscape in terms of labour processes and underlying economic structures or forces. Whilst this chapter continues in the recognition of the importance of ideology, it considers this in the context of landscapes' landscape *production* (through "work"), rather than consumption (via representations). More than scenery to be looked upon, landscape is "rematerialized" and reconnected with the labour relations that make (and sustain) it as what it is. In turn, the fact that landscapes themselves "do work" is emphasised: they serve as stage for further action and make concrete a vision (or version) of how the world is that comes to be accepted as taken for granted, or naturalised (Mitchell, 2003b). Like the bings introduced above, cultural landscapes, as *substantive* entities, both illustrate and are deployed in support of a variety of economic and social interests.

Chapter Outline

This chapter sets out how scholars can account for landscape through a materially focused, often Marxist lens. It situates cultural landscape study within wider understandings of the importance of acknowledging the material (as opposed to purely representational) world; something especially prominent in the development of political ecologies of landscape. The chapter examines what a Marxist approach to the cultural landscape involves and elaborates the framework offered by geographer Don Mitchell for considering landscape as both "a work" and that which "does work", appearing simultaneously fixed and in process. We further expand on this approach via two examples. Firstly, the links between labour and capitalism as they transcend the "local" landscape are considered via Mitchell's work on the cemetery in Holtville, California. Secondly, expanding the relationship between capitalist landscape production and forms of racial exclusion, we engage scholarship examining urban "Chinatowns" as a landscape "type", and that examining the linkages between cultural landscape, memorialisation, and race in the American South.

A Material (Re-)turn in Cultural Landscape Studies

Landscape scholar Kenneth Olwig influentially argued in the mid-1990s that too much focus had been directed towards landscape's abstract theorisation in terms of cultural categories, representations, and meaning-making. The result, he felt, was that landscapes were too rarely treated as material entities and environments. Attempting to diversify understandings of the term landscape, Olwig argued that the changing meanings of landscape required explicit (re)connection to the material, historical contexts that produced them (e.g. Olwig, 2015). Through extensive philological (the study of language) and historical-geographical research, he demonstrated how older notions of landscape, existing alongside (even predating) more painterly or scenic associations (see Chapter 2), emphasised it as specifically material: "an area carved out by axe and plough" (Olwig, 1993: 311). Within such understandings, landscape could not be separated from material conditions (and their organisation by specific communities through systems of ownership or law). Rather, it was something built up over time, through the actions – the labour – of those existing upon and through the land. In his words, landscape is named "a nexus of community, justice, nature, and environmental equity, a contested territory that is as pertinent today as was when the term entered the modern English language at the end of the sixteenth century" (Olwig, 1996: 630–631).

In the context of examining how the political-economic and social changes that characterised the Renaissance and emergence of capitalism unfolded, Olwig emphasises the value of understanding landscapes as "the world we are living *in*", not just "a scene we are looking *at*" (Wylie, 2007: 15). The transformation of the landscape to suit the ends of economic production materialises a particular economic ideology into the world, shaping in turn how the environment is both perceived and used. Defining landscape solely in terms of painting, scenery, or aesthetic conventions threatens to "obscure the relation between the aesthetic form of landscape and its substantive content" (Olwig, 1996: 631). In other words, we focus so much on the painted image that we potentially forget about the active lives and labours that literally make it what it is and take place within it.

Prompted by attention to the etymology (linguistic origins) of "landscape", Olwig promoted a shift in attention to landscape as both a political and material entity that became integral to several studies through (and beyond) the 1990s. In part, this was a reaction to a perceived overemphasis in subjects such as cultural geography upon representational politics that had tended to ignore the "matter" of nature (see Fitzsimmons, 1989) and also a development of debates carried over from the emergence of a more political engagement with ecological questions in the context of development during the previous decade (see Box 3.1). For Olwig, his work comprised a return to the "substantive nature of landscape", and an attention to what was "real rather than apparent" (Olwig, 1996: 645).

An attention to the relationship between material environments and ideology had also been proposed within Marxist spatial philosophy, particularly by theorist Henri Lefebvre whose work *The Production of Space*, published in 1974 in French, was made more widely available via an English translation in 1991. Lefebvre offered a powerful argument that space's materiality – its solidity, tangibility, and material properties or characteristics – was fundamental to appreciating its production within certain social contexts as the domain for existence, and as something which reinforced and served the interests and narratives of the powerful or ruling class. As he notably questioned:

> What is an ideology without a space to which it refers, a space which it describes, whose vocabulary and links it makes use of, and whose code it embodies? What would remain of religious ideology [...] if it were not based on places and their names: church, confessional, alter, sanctuary, tabernacle? What would remain of the Church if there were no churches? The Christian ideology [...] has created the spaces which guarantees that it endures.
>
> (Lefebvre, 1991: 44)

In short, work on representations had made clear that the idea of landscape was culturally specific and ideological (see Chapter 2). An attention to the material spaces that gave ideology form, made it mean something, and sustained its meaning over time was required. For geographer Don Mitchell, whose extensive body of work showcases such a materially focused interpretation of cultural landscape formation, failing to attend to the "stuff" of ideology meant one risked a "spiral [...] into untenable idealism" (Mitchell, 1996: 5), debating values and disembodied meaning-systems apart from any actual lived existence or the struggles defining it. For studies of landscape, "the relationship between material form and ideological representation" had to be the central question (Mitchell, 1996: 5) from which followed the matter of how a given scene was manifest, maintained, and made to seem natural. To do so, Mitchell asserted, required examining the *histories* of landscape production: not merely those of its representational production (art history) but labour histories and vernacular social histories so often obscured by an attention to elite ways of seeing. In this way, a sense of landscape's ideological *materiality*, as Lefebvre conceived it, reflecting a contested, political history, took centre-stage with Mitchell's critical Marxist landscape scholarship, as had come to be the case more broadly in the context of several social science works on environment and development.

Box 3.1 Political Ecologies of Landscape

Calls to "re-materialise" landscape study were indebted to and influenced by the emergent field of *political ecology*. This work came to prominence in response to the apparent apolitical treatment of environmental issues in development debates during the 1970s and 1980s. In particular, critical Marxist scholars objected to a narrative of "underdevelopment" appealing to notions of "natural limits" when explaining observed land degradation or agricultural unproductivity, often in the Global South (see Harvey, 1974). Political ecology offered critical corrective, reframing environmental challenges as the outcome of unequal power relations and/or specific political and economic conditions. Key studies – such as Piers Blaikie and Harold Brookfield's collection *Land Degradation and Society* (1987) – set out how issues such as desertification, soil erosion, and flooding could result from failures in land management connected to political and economic structures. Although not a term used to focus some of these earlier studies, landscape became a key concept in political ecology work as developed over subsequent decades (especially in recent work on conservation and its landscape frameworks), framing discussions of the social production of environments and ecologies, their contested meaning or interpretation by different groups, and an attention to the histories and legacies of colonialism underlying different landscapes (Neuman, 2011; also, Chapter 6). In the work of Neil Smith, emphasis was put upon the ways capitalist economies, ceaselessly searching for resources and wealth, "[transform] the shape of the entire world" leaving "[no] God-given stone […] unturned, no original relation with nature unaltered, no living thing unaffected" (2010: 8). Attempting to align the material form and processes of landscapes with economic (human) interests, the resulting proliferation of what Smith termed "second nature" reworked ecological systems into "social matter" with allocated value (2010: 65–68).

Framing the relationship between environment and capitalism in critical Marxist terms understands material landscapes and resources as "historically embedded and shaped by ideology and social relations" (Turner, 2016: 185). Attention likewise turned to consider the role of institutions like the World Bank and International Monetary Fund, especially in a global development context; as well as the unequal social and economic relations, territorial boundaries and borders, and character of resources themselves (including the ease with which they could be extracted, transported, or sold). These questions continue to be explored with reference to current environmental issues. Sociologist Jason Moore (2016) characterises the epoch of "the Anthropocene" as one in which the forces of capitalism have entangled as never before with ecological processes to produce "cheap nature" for human consumption (plantation forests, agricultural landscapes, new strains of GM crops that can be easily farmed, etc.). Such interventions often create unforeseen issues like the spread of disease through monoculture crops, soil erosion, or waterway pollution; and there is a need to comprehend how capitalism "creates new barriers to its own future" as it unleashes novel hazards or forces that potentially threaten future survival (Smith, 2010: 84). Elsewhere, efforts to incorporate landscape's experiential or "more-than-human" dimensions (see Chapter 10) within political ecology have been deployed to attend to how lived, site-specific histories of human–nonhuman interactions and land use produce particular contexts within which efforts to protect or manage ecologies (e.g. in conservation) are negotiated alongside the needs of politically or economically marginalised communities (see Barua, 2014; Margulies, 2019). Political ecology, then, remains a lively field of landscape study.

A Marxist Landscape Approach

Marxist approaches – drawing on the writings of Karl Marx and those who have sought to apply and recontextualise his ideas – have proven an influential means of "rematerialising" the cultural landscape. Within Marxist theory, the components of society (its institutions, spaces, social relations, values), termed its "superstructure", require understanding and explanation with reference to the economic mode(s) of production (and their associated logics) that comprise its "base". Whether organised around feudalism, an economy of slave-trading, or free market capitalism, everything in society is explainable with reference to the foundational structures of the political economy. In turn, questions like "[w]hy does the landscape look like it does" or "who made it look that way" (Mitchell, 1996: 5–6), approached on this basis, can reveal how the organisation of physical environments relates to practices and relations of labour.

Importantly, a Marxist perspective does not *necessarily* involve attention to the physical *land* or labour practices. John Berger's approach to art criticism (see Chapter 2) was avowedly Marxist, elaborating how representations of "landscape" (or other subject matter) glossed unjust systems of exploitation by celebrating or naturalising elite landownership. Other works focussing on the iconography or textuality of landscape advance similar arguments. However, Mitchell argues, such work risks treating "culture" (and by extension, cultural landscapes) in ways neglecting attention to material lived realities. References to "culture" made as though it is an entity with causal force, circulating through representations (or associated acts of categorising, valuing, or labelling), forget that culture is *itself* an ideological label applied to specific objects or activities and according them meaning. The things, to which culture refers, however, are themselves products of labour. Thus, the struggle over their meaning and significance (as "culture" or not) reflects the struggled-over nature of the social itself as an outcome of political-economic relations (see Mitchell, 1995). A Marxist stance on landscape foregrounds this struggle in its materiality. As Mitchell writes:

> as a material "thing" and a representation (no matter how abstracted) of that thing, landscape structures relations of labour. It is [...therefore] a powerful visual ideology; but it is also a structured portion of the earth within which people work and live and sleep, eat, make love, and struggle over the conditions of their existence.
>
> (1996: 2)

For the Marxist scholar of landscape, studies seek to reconnect the physical environment with the history and struggles of human labour. In turn, one might understand how, in service to powerful interests, landscapes are maintained in one or other form, as the work going into their making gets obscured or hidden. We turn to elaborate Mitchell's specific conceptualising of "landscape as labour" below.

Landscape and Labour

A focus on reading or decoding depictions of landscape (and their consumption) for what they reveal of the ideologies embedded through their representation only tells part of the story and too little regarding their physical production. After all, it is in its physicality that landscape exists as a "substantive reality, a lived place, a world produced and transformed, a commingling of nature and society that is struggled over and in" (Mitchell, 2003a: 792). Here, we elaborate how this physical environment can be examined as a question of *labour*: something worked on and in, and that emerges as a product of work and struggle. This "double reading", drawn from Mitchell, keeps hold of how landscapes are made and how they work.

Landscapes as Work

Mitchell argued that to understand landscape one needed to appreciate the work that produced it. Here "work" is not to be conflated with "employment" status or as a synonym for the "9 to 5". Rather, the term encapsulates the effort and time spent to organise or transform – to make – a material environment into the form it exists in at a given time. Such work can be human but need not be exclusively so. For example, a city park is the outcome of both nonhuman (we might say "natural") forces like erosion, weather, vegetation growth, or animal activity. It is also made through the actions of people: landscape designers and labourers; employed (or perhaps volunteer) gardeners maintaining an appearance; visitors using the park for leisure; the economic activities of a person or organisation who purchased the land and funds (or benefits from) such activities. All these elements, then, work towards the creation of "the park" and such work, or rather *labour*, is what creates (and sustains) any "landscape".

Taking this approach to landscape study, argues Mitchell, acknowledges both the historic labour that made a landscape and those labours ongoing in the present. This is not just the "living, form-giving fire" of contemporary action, but also the labour which is "dead" (2007: 561) that produces landscape. This concept of "dead labour" is less an attempt to articulate the labour of the (now) deceased than to acknowledge the continuing influence of work previously done. We tend to imagine the work of the past as "complete" yet it is, in actual fact, ongoing as it shapes how landscape appears, or functions, in the present. Therefore, dead labour names the expended energy incorporated into a commodity – an object with a saleable value, whether a strawberry (Mitchell, 2003b) or a landscape – that is no longer obvious or necessarily visible, yet "congealed" (in Marx's terms – 1987: 176) in it. Landscape expresses such dead labour in its form, which when perceived as "natural", or taken for granted, serves to hide or veil the struggles (both material and social) that went into its creation.

A notion of landscape as "hiding" or veiling something connects this approach with the work of those scholars (e.g. Denis Cosgrove 1998) focusing upon representations and art in the context of social histories of landscape formation (see Chapter 2). Indeed, for Mitchell, romanticised depictions of landscape in cultural media often cover over the histories, politics, and inequalities that are entwined within such scenes to this day. However, Mitchell aims to recover the labours obscured or marginalised by such depictions, and their critical analysis. Seeking to "de-fetishize" landscape – to connect it back up to the processes and histories of labour and struggle that served to turn its "raw materials" into objects, services, or products with value – was therefore a way of answering the question posed by critical Marxist scholar Neil Smith: "how does the geographical configuration of the landscape contribute to the survival of capitalism?" (2010: 4).

The Work of Landscape

In framing landscape in terms of *labour* and its relations, one risks an account of landscape that positions it as the determined, or inevitable, outcome of underlying economic structures. Mitchell complicates this, however, by proposing a "double reading" of landscape in terms of *both* its production (making) and *re*production (continuation). In other words, how it is that landscapes are made and then kept in a particular state of being, rather than continuing to transform, as they are worked upon. In this way, landscape is both an outcome of work and its staging ground. As Mitchell writes, landscape is

> best seen as both a *work* (it is the product of human labour and thus encapsulates the dreams, desires and all the injustices of the social systems that make it), and

> as something that *does work* (it acts as a social agent in the further development of place).
>
> (Mitchell, 2000: 94)

In other words, the form of landscape – reflecting the history of how it had been worked, and efforts to maintain it in the face of forces and processes that might transform it further – can be understood in relation to an underlying mode of production and the interests associated with it. Landscape is *never finished*, even if it appears unchanging, but always being reproduced, often via struggle (e.g. between workers and landowners). As Mitchell puts it, as different "social groups", each "with differing access to power, financial and social resources, and ideological legitimacy" interact and come into conflict, "the form of the landscape is produced" (1994: 10).

However, landscape is made to *appear* static. This reflects both the work to hold it in shape and the efforts of "powerful social interests" who "[try] to represent the landscape as a fixed, total and naturalised form" (Mitchell, 1994: 10). Consequently, the "structured permanence" of landscape is "duplicitous" (Mitchell, 1996: 27), appearing natural whilst being actually ideological. By keeping the landscape in a certain form, associated political, economic, or social conditions characterising that landscape (as well as being served by, or drawing legitimacy via appeals to, it) are also reinforced. Thus, depictions of a romantic, rural idyll (see Chapter 2) both rely on and maintain the class relationships that take material form in the way certain agricultural landscapes are organised, managed, or owned. Appearing timeless, the landscape is cast as a fetishised object, set apart from all relations of production. In turn, "the key questions for researchers on landscape" involve excavating "what sort of relationships it incorporates but is reluctant to make readily apparent" (Mitchell, 2002: 385).

Pause for thought: *Can you think of any examples where a landscape you had taken previously for granted was cast in a new light? Has visiting a museum, reading a street plaque, or talking to someone led you to question why a landscape appears as it does? In what ways?*

To summarise: Mitchell's double reading of landscape asks us to address two key aspects. Firstly, one must understand the ways in which the landscape has been physically made through relations and histories of labour. This includes labours past (or "dead") yet which remain influential. Secondly, one should consider how the form of landscape as encountered (or maintained) enables the reproduction of particular socio-economic (e.g. capitalist) relationships. For Mitchell, the value of considering landscape in terms of this dialectic "is that it forces us to understand hidden as well as obvious worlds, material practices as well as ideological impositions" (2002: 385). Landscape is both formed from, and a projector for, those with the power to control its labour: it is "power materialised" (Mitchell,

2012: 397). Attending to the material ways in which power is exerted (as well as the unequal labour relations producing material forms) strengthens the means to understand landscape's effects and refocuses attention towards social and economic justice. The following examples serve to clarify how cultural Marxist understandings of landscape can direct attention to processes of capitalist circulation and incorporate attention to the politics of cultural identity.

Landscape and the Circulation of Capital

Researching landscape through an ontology of labour has often produced work which focuses on a particular locale. Mitchell, for example, has applied this Marxist landscape lens to fruit farms, mining towns, exclusive suburbs, and public parks. And yet, the example which follows stresses that landscape should not be understood as bounded in an isolated set of circumstance. Rather, "local struggles intersect with wider scale processes to determine the morphology of the circulation in transformation of landscape in a particular locale" (Mitchell, 2007: 561). Landscape is therefore understood as the outcome of wider processes of capitalist circulation.

Mitchell demonstrates this via a study of the town cemetery in Holtville, located in the Imperial Valley, California. Where temperatures are often over a hundred degrees Fahrenheit (c. 40°C), this landscape is marked by the presence of around 350 brick-sized graves. Inscribed with the pseudonyms John and Jane Doe, these are the graves of unidentified bodies recovered in the area after crossing from Mexico into the US, some ten miles away, and collectively tracing "the geography of death that is the US-Mexico border" (Mitchell, 2007: 563). Thus, the graveyard landscape (and souls interred there) materialises in place the broader system by which migrant bodies, agricultural labour, and ultimately capital circulate. The landscape is also an outcome of a host of other "labours": border policing, migration policy, cold storage, attempts to locate family members, undertaking, burial, and activism.

The deaths that have produced this landscape are symptomatic of policies that have hardened the border, with increased enforcement and surveillance resulting in illegal border crossings taking place across more difficult terrain marked by mountains, canyons, and deserts. At his time of writing, deaths were a daily occurrence. Meanwhile, for those who survived the journey, there is "a clear bodily cost [...] counted in increased assaults, rapes, robbery, severe dehydration, sub-lethal hypo- and hyperthermia [...] exposure, snake and insect bites, infections from crossing toxic canals, and other maladies" (Mitchell, 2007: 563). Mitchell argues that one might be tempted to view this death-defined landscape as the "antithesis of [...] productive labour" driving the US economy. Yet, it is, in fact, material product of the labour relationships supporting that economy. As he argues:

> The 400 or so yearly deaths along the US-Mexico border, the dozens of anonymous deaths each year [...] the maiming and injuring of so many more: these are not by-products or unfortunate accidents or a tragic waste. They are

> instead, the vital foundation upon which the US economy, and therefore the production and transformation of nature is now founded.
>
> (Mitchell, 2007: 563–564)

The bodies in the anonymous graves within this cemetery at Holtville are, in other words, representative of the "dead labour" that constitutes the landscapes of Californian agriculture: "labour dead in its most literal sense" (Mitchell, 2007: 564). In an agricultural system reliant upon seasonal influx of displaced foreign workers; with few protections, little capacity to resist poor pay or conditions, and being easily replaced, this cemetery landscape constitutes "a node in the network of relations of production and reproduction" stretching from landowners and employers in north and east California to villages in Mexico and Central America. The graveyard "is where migration and the flows of capital [...] become landscape: literally" (Mitchell, 2007: 570).

One cannot separate this place from the wider agri-economic system that keeps the majority of the US, and numerous other countries, in an abundance of food. As with Mitchell's (2003b) previous examination of fruit picking and the landscape of strawberry production in California, his work highlights for us how landscape, like the commodity itself (e.g. the strawberry on the supermarket shelf), is so often separated from the labour integral to its production. The deaths recorded in Holtville are part of the same flow of money and produce that keeps the rolling fruit fields of California appear as they do, and which carries food to your table. This understanding of cultural landscape, then, as a material, worked-upon, struggled-over environment, helps us understand the world as produced via labour. Most importantly, it does so in a way that connects the here and now to other places and times, as Mitchell writes:

> Whatever lure the local may hold for [researchers] it must become one of understanding how particular places, particular landscapes, fit into a larger, scalarly complex mosaic of landscapes that themselves are never "naively given", as Sauer (1925) once suggested. Instead they are actively produced and struggled over, and it is the politics of production and struggle that ought to engage landscape studies.
>
> (Mitchell, 2001: 270–271)

Racialised Landscapes of Labour

The economic relationships which (re)produce landscape can also be considered in terms of how they are textured by the politics of cultural identity and associated struggles. Connections between landscape and race (see also Chapter 6) are important when applying Mitchell's double reading of landscape in relation to labour (not least given the centrality of migrant, non-White labour in the example above) whether rural or urban. The history of landscape study is itself overwhelmingly white and male. A decolonising of reading, writing, and teaching practices requires scholars and students to explore alternative landscape narratives as well as the racialised nature of everyday landscapes. As Richard Schein (2009) suggests, via his own work on his hometown of Lexington, Indiana, simply confronting oneself with the conventional framing or story of a landscape to ask "what's missing?" can be powerful. In his example, the courthouse in the town centre was once the site of a slave market, and recognising this history of how the landscape was made and remade, with certain elements erased, prompts questions about its production and subsequent memorialisation.

Geographer Kay Anderson (1987) has examined the connections between race and labour in the context of her work on urban landscape, specifically her exploration of Vancouver's Chinatown. Arguing that the urban "China Town" constitutes a "western landscape type" (1987: 583) in which there is a stark difference between Chinese (nationality or culture) and Chinatown (space), she examines how designating space in terms of nationality, or indeed race, is a complex process serving numerous internal and external interests. In Anderson's example, Chinatown, as an area of Vancouver city (Figure 3.2), came to be defined by the efforts of a white, governing population to control and contain incoming migrant Chinese workers around the turn of the nineteenth century. Understandings of race and space became intertwined within this landscape as issues of sanitation and vice (prostitution) within Chinatown served to reinforce and institutionalise racial categorisations. The Chinatown that Anderson describes is therefore a product "of work"; both

Figure 3.2 Chinatown Gate, Vancouver, BC. Image by Suman Chakrabarti. Wikimedia Commons, CC BY 2.0.

in terms of external, racist exclusion, violence and marginalisation, and the lived experiences of resident Chinese immigrants who were "active agents in their own 'place making'" (1987: 583). Today, the Chinatown landscape is taken for granted as a tourist attraction and integral part of the city: naturalised and uncontentious. The outcome of historic struggles and exclusions concerning immigrant labour, it continues to "do work" in the service of tourism.

Work such as Anderson's highlights that the materiality of landscape is not only an outcome of economic relationships but also of associated social dynamics. The landscape that results is therefore as much racial "discourse materialized" (Schein, 2009: 395) as it is the history of labour struggle. In a similar vein, Perry Carter, David Butler and Owen Dwyer's (2011) work seeking to "defetishize the Plantation" explores connections between race and labour in the context of memorial work in the Southern US. As they discuss, plantation museum narratives in the American South have often fetishised the history of their creation by not addressing the role (and labour) of enslaved peoples in producing them. As the authors emphasise, through this process of racial erasure, the plantation landscape, as the products of human labour, appears seemingly from nowhere, "eliding the narratives of the enslaved labor that made such edifices possible". As they elaborate:

> In this sense plantation museums are recycled landscapes, which posthumously extract wealth from the labours of former slaves. Hence, the place of the enslaved on and through the landscape is doubly obscured.
>
> (Carter et al, 2011: 129)

Plantation museums, Carter et al argue, have served as "a particular type of fetish commodity" (2011: 129) by which the landscape is used to stage the selling of a particular history, privileging "big house" narratives over the lives of the enslaved peoples labouring in the background of such tales. This "story of the powerful" obscures the heritage of those whose physical labour actually made the landscape as part of a political economy based in slave-owning.

In the same way that others defetishise the strawberry on our plates (Mitchell, 2003b), the plants that adorn our homes (see Box 3.2), or the stereotypical tourist scene of "Chinatown" (Anderson, 1987), Carter et al's scholarship provides a challenge to museum narratives by "[r]eintroducing the enslaved into the public history of the Southern plantation" (2011: 129) and reconnecting these landscapes, as material environs, with the work and struggle that created them. Such a process is both practical and ideological, requiring the production of new resources from extensive research, informing revised tours and exhibitions. Yet, as eighteenth-century abolitionists defetishised sugar by highlighting its production via the labour of enslaved peoples to end such human subjugation, so can attention to labour *do work*. Shifting the narratives of this landscape of the American South, and the material forms through which that narrative is dispersed (plantation museums), involves reimagining that landscape in more just, equitable, and accountable ways. This new landscape narrative has an audience seeking more than clichéd depictions of Southern elegance, and as the landscape is put to work in new ways, we are reminded that it is always in process.

Box 3.2 Labouring for a Beautiful Landscape

Cultural landscape studies have scrutinised the role of representation, labour, and social values in the creation of aesthetically pleasing landscapes. At the same time, research has acknowledged the work such landscapes do to reinforce social norms (see Duncan and Duncan, 2003 on suburban gardens). Emma Crane's examination of "beautiful landscapes" offers such an analysis, considering the labour that produces home or garden plants via the nurseries in which they are grown and lives of those labouring for their cultivation. As she outlines, a plentiful, marginalised, and subsequently exploited Indigenous migrant labour which sustains the production of houseplants across the US, with most of this "skilled", "arduous", and "dangerous" work done by women and children.

Her research focuses on the agricultural suburb of Homestead in Greater Miami, Florida. Crane reframes the suburb as neither the quaint nor safe landscape of "ticky-tacky" boxes (to evoke Malvina Reynolds' lyric). Instead, as Crane (2023: 211) describes, "the plant nurseries of Homestead are one lucrative aftermath of ongoing, transnational war-making" as "[c]ycles of war and displacement in Guatemala render Maya people available as displaced wage laborers" consequently led to work in plant nurseries in the US. Whilst we may be more accustomed to think of the detention camp or military base as militarised landscapes, Crane's example demonstrates how other, more mundane landscapes are likewise "sustained through militarism" and security practices as they create the conditions for their ongoing reproduction. Conflict, genocide, and, latterly, "peacetime" dispossession produce a labour source for plant nurseries who are limited by the "surveillance and policing" that they are subject to in their mobility and capacity to find other kinds of work (Crane, 2023: 217).

Crane therefore highlights how the labour involved in the creation of a houseplant includes not only the physical labour of cultivating the product, but the effort expended to maintain the landscape in a given form such that it continues to serve as a node in a web of economic relations. These connections conjoin the plant-as-commodity to places and processes far beyond the home in which it now sits, or indeed the nursery where it was grown and sold. The beautification of the suburban landscape is thus understood as the "lush aftermath" of military activities and dispossession abroad, offering a reminder that studies of landscape must be "attentive to the constitutive relationship between dispossession and accumulation" (Crane, 2023: 231) as these are bound up with both racial inequality and geopolitical dynamics, echoing into everyday environments.

Chapter Summary

Box 3.3 Key Concepts

Chapter 3

Material(ism)
Marxism
Labour
Circulation of capital
Landscape as work

Understood as material entities reflecting the labour that made them, cultural landscapes can teach us much about the past and ongoing struggles resulting from unequal economic relationships (Box 3.3). Where a landscape might *appear* to be static or "finished", Marxist approaches prompt us to draw back the veil of appearance and taken-for-granted history to interrogate their (re)production. Thus, a material-Marxist approach examines the lived, material reality that representations of landscape (see Chapter 2) serve to obscure, emphasising its manifestation through struggle in service to a ruling ideological interest. As Mitchell writes,

> [s]ocial struggle makes the landscape, and the landscape is always in a state of becoming; it is never *entirely* stable. Yet landscape is also a totality. That is, powerful social actors, as we have already suggested, are continually trying to represent the landscape as a fixed, total, and neutralized entity – as a unitary thing. Landscape is thus best understood as a kind of produced, lived, and represented space constructed out of the struggles, compromised, and temporarily settled relations of competing and cooperating social actors: it is both a thing (or suite of things) […] and a social process, at once solidly material and ever changing.
> (1996: 30)

This Marxist account of landscape foregrounds its materiality and encourages a grounded, morphologically attentive, ideological critique. In this way, it connects with other approaches discussed elsewhere in the textbook that seek to understand how landscape is made or produced, and the purposes this serves in relation to the exercising of power. Therefore, this chapter makes clear the value of understanding the processes by which landscape is made and maintained, and the importance of not assuming its present constitution to be simply "natural".

Suggested Seminar Activities

Activity 1: Read Kay Anderson's (1987) piece "The Idea of Chinatown" and consider (or discuss) the following questions:

- What sort of labour has created the landscape of Vancouver's Chinatown? Who is undertaking that labour and when was it done?
- In what ways does this landscape *continue* to do work that serves particular economic (or social) interests?

Given this article was written several decades ago, in what ways does it continue to be relevant for understanding relationships between race, labour, and the urban landscape today? Are any aspects of the article no longer relevant?

Activity 2: Read Madeleine Eriksson and Aine Tollefsen's 2018 piece on the Swedish berry industry and discuss in class.

Consider the following questions:

- How are landscapes and landscape representations altered by migrant labour?
- In what ways do "ideologies of race, gender and class … become part of the landscape" (2018: 79)?

Next, try to make links between the reading and other examples:

- Mind-map other products which "remake geographies and form landscapes" (2018: 80). You could consider popular foodstuffs or the most common superfoods. What are the landscapes of labour involved in their production and consumption? This could involve researching and giving a short presentation on the links between landscapes and labour which are embodied in your chosen product.

Recommended Reading

Carter, P., Butler, D., and Dwyer, O. 2011. Defetishizing the plantation: African Americans in the memorialized south. *Historical Geography* 39: 128–146.

Crane, E. 2023. Lush aftermath: Race, labour, and landscape in the suburb. *Environment and Planning D: Society and Space* 41: 210–230.

Mitchell, D. 1996. *The Lie of the Land: Migrant Workers and the California Landscape.* Minneapolis: University of Minnesota Press.

Mitchell, D. 2003b. California living, California dying: Dead labour and the political economy of landscape. In Anderson, K., Pile, S., and Thrift, N. Eds. *Handbook of Cultural Geography*. London: Sage, pp. 233–248.

References

Anderson, K. 1987. The idea of Chinatown: The power of place and institutional practice in the making of a racial category. *Annals of the Association of American Geographers* 77: 580–598.

Barua, M. 2014. Bio-geo-graphy: Landscape, dwelling, and the political ecology of human-elephant relations. *Environment and Planning D: Society and Space* 32(5): 915–934.

Blaikie, P. and Brookfield, H. Eds. 1987. *Land Degradation and Society*. Abingdon: Routledge.

Carter, P., Butler, D., and Dwyer, O. 2011. Defetishizing the plantation: African Americans in the memorialized south. *Historical Geography* 39: 128–146.

Cosgrove, D. 1998. *Social Formation & Symbolic Landscape*. London: University of Wisconsin Press.

Crane, E. 2023. Lush aftermath: Race, labour, and landscape in the suburb. *Society and Space D* 41(2): 210–230.

Duncan, J. and Duncan, N. 2003. Can't live with them; can't landscape without them: Racism and the pastoral aesthetic in suburban New York. *Landscape Journal* 22(2): 88–98.

Eriksson, E. and Tollefsen, A. 2018. The production of the rural landscape and its labour: The development of supply chain capitalism in the Swedish berry industry. *Bulletin of Geography: Socio-Economic Series* 40: 69–82.

Fitzsimmons, M. 1989. The matter of nature. *Antipode* 21(2): 106–120.

Flyn. C. 2021 Islands of Abandonment. London: William Collings.

Harvey, D. 1974. Population, resources and the ideology of science. *Economic Geography* 50(3): 256–277.

Harvie, B. 2004. The Mechanisms and Processes of Vegetation Dynamics on Oil-Shale Spoil Bings in West Lothian, Scotland. PhD Thesis, The University of Edinburgh.

Lefebvre, H. 1991. *The Production of Space in Critical Social Theory*. London: Verso.

Margulies, J.D. 2019. Making the 'man-eater': Tiger conservation as necropolitics. *Political Geography* 69: 150–161.

Marx, K. 1987. *Capital, Vol. 1*. New York: International Publishers.

Mitchell, D. 1994. Landscape and surplus value: The making of the ordinary in Brentwood, CA. *Environment and Planning D: Society and Space* 12: 7–30.

Mitchell, D. 1995. There's no such thing as culture: Towards a reconceptualization of the idea of culture in geography. *Institute of the British Geographers* 20: 102–116.

Mitchell, D. 1996. *The Lie of the Land: Migrant Workers and the California Landscape*. Minneapolis: University of Minnesota Press.

Mitchell, D. 2000. *Cultural Geography: A Critical Introduction*. Oxford: Basil Blackwell.

Mitchell, D. 2001. The lure of the local: Landscape studies at the end of a troubled century. *Progress in Human Geography* 25: 269–281.

Mitchell, D. 2002. Cultural landscapes: The dialectical landscape – Recent landscape research in human geography. *Progress in Human Geography* 26: 381–389.

Mitchell, D. 2003a. Cultural landscape: Just landscapes or landscapes of justices? *Progress in Human Geography* 27: 787–796.

Mitchell, D. 2003b. California living, California dying: Dead labour and the political economy of landscape. In: Anderson, K., Pile, S., and Thrift, N. Eds. *Handbook of Cultural Geography*. London: Sage, pp. 233–248.

Mitchell, D. 2007. Work, struggle, death, and geographies of justice: The transformation of landscape in and beyond California's imperial valley. *Landscape Research* 32: 559–577.

Mitchell, D. 2012. *They Saved the Crops: Labor, Landscape, and the Struggle over Industrial Farming in Bracero-Era California*. University of Georgia Press: Athens, GA.

Moore, J. 2016. *Capitalism in the Web of Life: Ecology and the Accumulation of Capital*. London: Verso.

Neumann, R.P. 2011. Political ecology III: Theorizing landscape. *Progress in Human Geography* 35(6): 843–850.

Olwig, K. 1993. Sexual cosmology: Nation and landscape at the conceptual interstices of nature and culture, or: what does landscape really mean? In Bender, B. Ed. *Landscape: Politics and Perspectives*. Providence: Oxford, pp. 307–343.

Olwig, K. 1996. Recovering the substantive nature of landscape. *Annals of the Association of American Geographers* 84: 630–653.

Olwig, K. 2015. Nationalist heritage, sublime affect and the anomalous Icelandic landscape concept. *Norsk Geografisk Tidsskrift – Norwegian Journal of Geography* 69(5): 277–287.

Richardson. C. (2012) 'Waste to Monument: John Latham's Niddrie Woman: Art & Environment', in Tate Papers no.17, www.tate.org.uk/research/tate-papers/17/waste-to-monument-john-lathams-niddrie-woman, accessed 21 August 2025.

Schein, R.H. 2009. A methodological framework for interpreting ordinary landscapes: Lexington, Kentucky's courthouse square. *Geographical Review* 99(3): 377–402.

Smith, N. 2010. *Uneven Development: Nature, Capital and the Production of Space.* London: Verso.

Turner, M. 2016. Ecology: Natural and political. In: Castree, N., Demeritt, D., Liverman, D. and Rhoads, B. Eds. *A Companion to Environmental Geography.* Chichester: Wiley-Blackwell, pp. 181–197.

"What's shale?". n.d. Shale Trail website. Available at: www.shaletrail.co.uk/whats-shale/isl and-refuges/ [Accessed February, 2022].

Wylie, J. 2007. *Landscape.* Abingdon: Routledge.

4 The Landscape as a Text

Introduction: Reading the Landscape

Niagara Falls, Ontario, Canada (Figure 4.1) is, as Patrick McGreevy argues, a landscape offered up to visitors as "a kind of text to be interpreted" (1992: 50). He describes how, for tourists and travellers in the mid-nineteenth century, the sight of Niagara's raging torrents; its abyssal drop into a churning plunge pool; and the rising clouds of vapour were understood almost as a written document, encoded with meanings and references that "related directly or indirectly to the symbolism of death" (50). But how did this landscape come to be so forcefully attached to – or inscribed with – these ideas?

For McGreevy, travellers and visitors to this site "who articulated the metaphorical meaning of Niagara's landscape" produced "written texts" expressing ways of understanding the scene (1992: 72). As the idea of death was romanticised through gothic art and literature in the eighteenth and nineteenth centuries, so a set of defined, shared readings was further established with regard to the Falls. In writing about Niagara, the landscape became a metaphor for the process of dying itself. The brink symbolised the border between life and death; the plunge, the fall into the afterlife; the abyss, the unfathomable and indiscernible character of death; and the rising mist and rainbow, imbuing the site with an almost religious feel, offered a glimpse of heaven, and the possibility of peace.

Informing, reinforcing, and being reinforced by such acts of writing/reading the landscape were the material facts of the Falls as a place of death(s). Hundreds of bodies have been pulled from its waters since the eighteenth century – something enthusiastically detailed by guidebooks from as early as the 1830s (McGreevy, 1992). That association with death has seen emerge a variety of horror-based museums and attractions nearby. The idea that the Falls connected "real danger" to life was enhanced by an "accumulation of stories" showing how this place might confront a visitor with their mortality; stories historically reproduced through personal and communal acts. Suicide attempts led to further "inscriptions" upon the landscape, such as fences, notices, and crisis phones. So too are efforts to *defy* death at the Falls: daring thrill seekers ridden the rapids in boats and even taken the plunge in barrels. In 1957, a religious leader attempted to walk on Niagara's waters before being swept to his demise. And the site has even hosted peace conferences

DOI: 10.4324/9781003172451-5

Figure 4.1 The Niagara River seen from Skylon Tower in Niagara Falls, Ontario, CA. Image by Nandaro. Wikimedia Commons, CC BY-SA 3.0.

(evoking the peace that follows the dealing of death) during the American Civil War and featured as a potential early contender for siting the UN.

This example demonstrates the importance of considering the intertwined cultural relationships of written texts and material landscapes as the environment is framed, written on, encoded with meaning, or subject to different, overlapping readings. It prompts consideration of the ways in which studies of landscapes might borrow conceptual and practical approaches from the study of texts. The Falls offer "a screen" onto which various cultural ideas are projected. They also possess a particular material character (reworked in a variety of ways) that affords certain interpretations. Therefore, this chapter examines the idea of the landscape as a text. Doing so, it provides a bridge between preceding and following chapters. Whilst Chapter 3 outlined a way of examining landscape informed by a structuralist, critical Marxist understanding of how cultural-environmental relations express underlying economic processes, here landscape is considered as textual entity in ways that move from structuralist towards *post-structuralist* understandings. Rejecting explanations derived (solely) from underlying structures, such work recognises the *multiple*, contested ways in which cultural landscapes are (re-)written and reinterpreted.

Chapter Outline

The chapter considers the "text" metaphor in studies of cultural landscapes as a long-established and enduring way of describing human relations with the environment. It discusses how landscape studies have drawn upon semiotics – discussed here with reference to the work of Roland Barthes – to understand landscapes as "written" or "read" by different communities. As the idea of "landscape as text" has come to characterise a turn from structuralist to post-structuralist thinking, the chapter considers the influence of this kind of philosophical thinking upon landscape scholarship. The chapter thus explores a cultural landscape, encoded with signs, which can thereby be understood as a "meaning-making device" (Lindstrom et al, 2018) with multiple interpretations which exist, jostle, and mingle. The chapter ends by examining the relationship between specific written texts and our practical encounters with landscape.

Reading the Landscape

Demonstrably, "reading" landscape has long been central to its study. Framing landscapes as "texts" precedes the relatively recent efforts to explicitly conceptualise landscape in relation to post-structural theories of "the text" (see Daniels and Cosgrove, 1993). Indeed, since its beginnings, the cultivation of a "morphologic eye" was figured essential to any successful differentiating or "reading" of cultural traces when considering the cultural landscape (Sauer, 1956: 290). The Berkley School approach (Chapter 1) presented looking at landscape as a key part of understanding it as the "literate" scholar drew "selective observation and critical judgement" to propose connections between form and process (Sauer, 1956: 295). Thus, Sauer and his adherents understood landscapes as *palimpsests*: "documents partially erased and overlain with newer forms and patterns", containing a wealth of information and clues to their histories (Duncan and Duncan, 2010: 228).

An entity that communicated information about human activities, landscapes could be "read" for evidence of cultural or social change. Our surroundings were in fact the cultural "spoor", or trail, left in the wake of such historical transformation. Early work in this vein tended towards what Jim Duncan terms "object fetishism": the observation of environmental features and cultural artefacts (e.g. building styles) revealing, to the trained eye, a "chronology of events and cultural value systems" (Duncan, 1990: 12). This approach is exemplified within Donald Meinig's *The Interpretation of Ordinary Landscapes* (1979), in which essays by various authors argued for sustained attention to ordinary, everyday environments so as to uncover the values, tastes, or dominant narratives characterising society. For instance, American geographer Pierce Lewis describes landscape in his essay as society's "unwitting autobiography" (1979: 12). The careful study of this "text" – which could be read as the accumulation of our tastes, values, dreams, and beliefs – by a series of prescribed "axioms" would reveal core truths about human society. Indeed, he argued that this "cultural record" (including everything from architecture to billboards to garden furniture) was "liable to be more truthful" by virtue of being set down in an "unthinking" way (1979: 12). This did not mean landscapes were *easy* to read – as messy documents, overlain with multiple traces ("palimpsests") they took time to disentangle – but rather that, by way of a persistent attention to "ordinary" scenes, we might "raise questions we had not asked before" (Lewis, 1979: 27).

> ***Pause for thought***: *Can you list some examples of objects in the landscape that might be read in this way as evidence for different cultural values?*

Meinig and others proposed reading the cultural landscape to aid an understanding of how certain landscape "types" emerged or were coded with social significance. The "suburb", "main street", or "village" represented distinctive *kinds* of landscape; their objects and arrangements betraying the characteristics of inhabiting communities and made legible through (broadly a-critical) reference to diverse materials including documentary records, fiction, images as diverse as artworks or postcards, and popular representations that, today, might include social media (Duncan and Duncan, 2010) that served as a record of environmental content. As Karen Morin summarises:

> Just as a book (text) is made up of words and sentences arranged in a particular order with meanings that we read, so landscape has elements arranged in a particular order that we can translate into language, grasp meaning and read.
> (2009: 289)

Language and Landscape

Throughout the 1980s, as studies of landscape were influenced by a range of theoretical and methodological insights from across the humanities, philosophy, and cultural studies (see Chapter 2), a new sense of how they might be read took hold. A reading focused on morphology gave way to embracing questions of *interpretation* and symbolic meaning. Material or represented landscapes were "configurations of symbols and signs" (Cosgrove and Jackson, 1987: 96) that geographers or literary theorists (amongst others) could use to explore the relationship between *language* and environment. Moreover, straightforward notions of "reading" ones' surroundings were complicated by theoretical discussions about how language worked to generate and communicate meaning.

In geography, for example, humanist approaches to space and place examined the value of literature and written texts beyond being mere records of landscape character or change (see Pocock, 1981). Instead, literature could reveal aspects of the human experience of landscape. For example, Yi-Fu Tuan's writing on space and place often calls upon literature to explain the significance of certain environments and experiences of place. He demonstrates how landscape "types" (such as rainforests or deserts) became established objects of cultural imagination or fascination, their extreme or challenging nature revealing the ways in which encounters with these environments become charged with "what human beings truly fear and desire" (Tuan, 2013: 173).

Whilst such engagements with literature have been criticised for an inward focus and appeals to essentialised "human nature" or spatial experience, the work of Tuan and other humanist scholars remains significant. It foregrounds the power of *language* in shaping perceptions of the world – something we discuss further below.

The Structure of the Text

The metaphor of landscape as text provided an invitation to apply the tools of literary criticism to a range of environments (Duncan and Duncan, 1988). Influenced by these approaches, scholars subsequently turned to interrogate how landscapes were constructed, and read, via their symbols or icons. Semiotics offered scholars of landscape one framework for interrogating the world as a text, understood as a composition of objects that conveyed meaning both individually and by virtue of an arrangement in relation to each other. Ferdinand de Saussure is often credited as the originator of semiotic analysis, but his largely structuralist ideas were extensively critiqued and expanded over the twentieth century, not least in the context of post-structuralist theorising that challenged any straightforward interpretation of meanings as fixed. Such a development is readily observable in the work of French philosopher Roland Barthes, whose thinking has been influential within cultural landscape studies (see Duncan and Duncan, 1992).

For Barthes, the semiotician (or "mythologist") decoding the construction and workings of cultural beliefs needed to interrogate not only written works. All manner of objects were considered "open to appropriation by society" through language, invested with meaning and serving as "signs" communicating it (Barthes, 1981: 109). The sign, as the base unit of semiotic analysis, is understood by Barthes (after Saussure) as the combination of "signifier" (an object, word, image, sound – the "thing" carrying meaning) and the "signified" (the meaning being conveyed). In the coming together of these components and creation of an object that is "read" for particular meaning(s), we have the "sign": "a type of social *usage* which is added to pure matter" (Barthes, 1981: 109).

Consequently, any sign (e.g. an image of an apple that stands in for the real apples on sale at a grocer) could be further imbued with additional meanings, made understandable within a given geographical, historical, or cultural context. For example, in a Western, Christian nation, an apple placed next to an image of a naked human or a snake becomes a symbol of *knowledge* (evoking Genesis in the Bible). Thus, a second-order semiological chain is established: the sign (apple) becomes available anew, carrying a more abstract meaning and ready for (re)inclusion within other compositions of signs. For Barthes, this process of abstraction and the layering of higher-order meanings constitutes how cultural "mythologies" emerge. The original history – indeed, material reality – of a first-order sign retreats as it gains new associations. He proposed to examine how everyday objects came to hold cultural significance with reference to a range of underlying rules or structures. Layered atop one another, meanings attached to things become part of a vast system of values or relations between objects, qualities, and ideas.

In one essay, Barthes describes the way in which guidebooks choreograph encounters with unfamiliar environments, turning the content of places into symbols that represent abstract ideas like the rural idyll or exotic other. For example, in one guidebook's description of the Spanish landscape, it is represented not as a lived environment, but a scene populated by variously iconic signs of "Spanishness". The result is that the actual land and inhabitants become merely representatives in "a vast classical ballet" of idealised qualities that "serves to mask the real spectacle of conditions, class and professions" (Barthes, 1981: 75). The homes, town squares, and people themselves are mythologised for visitor consumption and the landscape becomes "an uninhabited world of monuments" (Barthes, 1981: 76), rich in history yet with its present removed. We can recognise how, today, contemporary tourist landscapes are similarly textualised by guidebooks and the proliferation of review websites, such as TripAdvisor. These platforms overlay and reframe specific locations with abstract values and promote specific readings of a given environment (see Bassett, 2023).

Within a semiotic approach, culture is understood as a system of signification with rules dictating how certain values are inscribed upon, or read off of, the landscape. Cultural landscapes are "sign systems", their constituent parts forming a whole in which "each of the elements is related to others and [...] combined into sequences according to certain codes" (Lindström et al, 2018: 76). In other words, landscapes are understood to have symbols and ideas materially written into their form that work to convey particular ideas. The landscape scholar must identify these signs and discern their meaning for the culture that inscribed them.

Towards Multiple Interpretations

That one can straightforwardly decode landscapes' meanings according to semiotic rules is a stance that can be robustly critiqued, not least from the outlook of post-structuralism. Highly influential upon humanities and social science research since the latter 1970s and 1980s, post-structuralism names a diverse range of conceptual approaches and lineages, key thinkers, and associated concepts – some of which feature here (and in Chapters 5 and 6). Whilst not any single project or aligned stance, crudely put the emphasis of post-structural scholarship has tended to be upon how particular ontologies (understandings of the world, what it comprises, how it is experienced or works) take shape and become stabilised, reproduced, and conceived as "true" (or a given) within particular relations of power. Attention is therefore also paid to how knowledge about the world is produced: to questions of epistemology (theories of knowledge) and the workings of different epistemological frameworks (the practices and standards for constructing knowledge).

When it comes to landscape studies, the implications of post-structuralism are neatly summarised by geographers Nancy and James Duncan:

> Post-structuralism takes very seriously the idea that a "view from nowhere" does not exist. In other words, no perspective is thoroughly objective; all perspectives are partial in significant and interesting ways. Given this belief, post-structural research on landscapes tends to focus on different readings or interpretations. These readings are not only individual, but can be understood as shaped by cultural discourses with multiple and complex histories that can only be very partially traced.
>
> (2010: 230)

The post-structuralist landscape scholar recognises there is no single reading of landscape or its meaning, but many. Different readings may coincide or conflict, and efforts are made by those in powerful positions to assert some readings and interpretations over others. Thus, particular readings of landscape are made more or less possible according to the wider assumptions, narratives, or frameworks that inform how landscapes are read in particular contexts, themselves the product of history. The landscape text, then, must be approached in ways enabling attention to questions of authorship, power, interpretive practice, and reception.

The idea of "the text" is also complicated and extended to encompass all manner of entities communicating meaning to those encountering them. As Marcus Doel argues, texts might name "anything that signifies something for someone or other" (2023: 245), including written documents, but also road signs, coloured wheelie bins, artworks, architectural styles, even the layout of an office or town square. Each work communicates or evokes meaning to those engaging them; each is differently "read". Likewise, landscapes can be seen as "signifying assemblages [collections of objects] that are like written texts, cut off from the intentions of their authors, and open to new interpretations by readers who become their authors" (Duncan and Duncan, 2010: 231). Approached as texts, landscapes appear "inherently dialogic" (Lindström et al, 2018: 85) and the landscape scholar must consider the interaction between writers, readers, and competing versions.

Re-reading the Landscape

Writing in the late 1980s, geographers Duncan and Duncan drew on cultural studies and literary theory to articulate how "[v]irtually any landscape can be analysed as a text in which social relations are inscribed" (1988: 123). The scholar of landscape, they reason, should attempt to read the text(s) of the landscape, understand its processes of "authorship", and grapple with its possible interpretations just as they might close-read any literary work or document. Such acts of interpretation are also political, with "material consequences" for how environments get understood and are acted on. The landscape is written with different interests in mind and read in different ways, by different agents, and its readings have implications for how people go about acting in or upon the world.

In Duncan and Duncan's framing of the landscape as a communicative device, several concepts came to the fore. First was the idea that each text (or landscape) had to be put into the context of other texts. It possessed an "intertextuality" since each text was always read (or re-read) through other texts coming before, or after, it; with readings that had implications for how meaning would be interpreted. As Duncan and Duncan (1988: 119) argue, "[a] text encourages the reader to carve it up, to rework it, to produce it [...] it is a space in which the reader as writer can wander". The same could be said of *landscapes* as material *and* representational systems of meaning-making and signification. The question, as geographer Richard Schein puts it, therefore becomes the following: how is this landscape, "this place", related or connected to landscapes and social processes in other places (1997: 662)?

Second, the role of discourse (see also Chapter 5) is to be examined. This concept articulates a broader framework composed of narratives, texts, values, and shared understandings, with implications for the interpretation of texts that are socially privileged, permitted, or even possible in particular contexts. The landscape-as-text idea retains a sense of landscape's *ideological* function since, where surroundings feel familiar and "unquestioned", their form viewed as "concrete evidence about how society *is* organised can easily become seen as evidence of how it *should*, or *must* be organised" (Duncan and Duncan, 1988: 123).

Important here is that interpretations of a landscape text, or its associated texts, are not infinite, but bounded by social and historical context. The reader sits amidst a set of social relationships, within which conventional readings are taken as a given. At the same time, the version of landscape (how it is materialised and read) that comes to the fore always remains "a question of the politics of interpretation" (Duncan and Duncan, 1988: 123), as well as how interpretations are promoted or discouraged. In practical terms, this approach to understanding landscapes' "textuality" requires us to examine how:

> through the vocabulary of various conventional forms – signs, symbols, icons, and specialised tropes in the landscape – people, particularly powerful people, tell morally charged stories about themselves, the social relations within their community, and their relations to a divine order.
>
> (Duncan, 1990: 20)

Box 4.1 The Urban Landscape Text

Urban landscapes are authored by various actors and read in conflicting ways as collections of signs. As spaces charged with emotion and politics (see Chapters 9 and 11), they play host to unfolding acts of "iconographical warfare [...] in the form of graffiti, murals and architectural constructions" (Yoon, 2020: 428). The following two examples from Southern Asia – one historical and one contemporary – illustrate how urban landscapes come to be written, interpreted, and contested in practice.

Example 1: The Last God-King of Kandy, Sri Lanka

In his study of the early-nineteenth-century urban landscape of Kandy, Sri Lanka – seat of a Buddhist kingdom on the island prior to its fall to colonising British forces in 1813 – James Duncan (1990) discusses how the last Kandyan King, Sri Vikrama, sought to strengthen and legitimise his rule through the city's material fabric and how it was "read". As Duncan describes, a tension between two competing discourses of kingship existed within Kandyan society. The *Asokan* discourse emphasised the benevolent role of the Buddhist king. The *Sakran* discourse, informed by the influence of Hinduism, positioned the king as himself a deity to be glorified via rituals, monuments, and the reconstruction of the urban landscape in ways "modelled upon textual descriptions of the cities of the gods in heaven" (Duncan, 1990: 40). Across Kandy's history, rulers were seen to lean towards one or other of these discourses, as well as attempting to combine or act in accordance with both. This history was palpable in the structure of Kandy's landscape, containing numerous "signs" that communicated the king's position and relationship to his subjects as religious written texts, and their differing interpretations "were transformed into landscape texts" (Duncan, 1990: 49) producing a different environmental form.

Sri Vikrama, taking the throne at a time when the court had become increasingly "Hinduizied" (see Duncan, 1990: 39, 196), leant more towards *Sakran* ideas of kingship. Facing rebellious feeling on the part of a Buddhist-leaning nobility in the wake of military victory over the British in 1803, he sought to reinforce his status via an extensive building programme that materialised references to the world of the gods, as well as key cosmological events and scenes from Kandyan creation myth, into the city's physical structure. Transforming the landscape to represent "the cosmos in miniaturized forms", the urban surroundings would project "the image of celestial space" to channel godly power and reinforce Vikrama's credibility (Duncan, 1990: 48–49). In one instance, the king conscripted peasant labour (something normally done only for civic or religious projects) to enlarge the city's decorative lake, renamed the "Ocean of Milk", after the primordial sea out of which the gods had created the world (Duncan, 1990: 97–101). This association, in the form and naming of the landscape, between the city and heaven; and thus the king and the gods, was enhanced further via other features such as a "wave swell wall" (Duncan, 1990: 101–107) evoking the churning seas of creation. These insertions aimed to encode the narrative that the king, inhabiting a landscape akin to that of the gods, was *himself* a god.

After the British captured the city in 1815, the landscape of the Kandyan highlands was further textualised through the romanticism evident in the writing and aesthetic conventions of those British and European travellers who came to consume its rugged, picturesque scenery (see Chapter 6). Indeed, the environment was transformed to accommodate such readings, with new buildings and planted vegetation evoking the style of the English countryside (see Duncan, 1998). Thus, from symbolising the domain of the gods, through its re-inscription in colonial travel writings as a familiar European countryside, the landscape of Kandy betrays the fact that, rather than one reading, many interpretations are possible.

Example 2: "Geomancy" and Banking, Hong Kong

Amongst several examples that demonstrate different urban landscapes as host to an ongoing "war of symbols", Hong-key Yoon (2020) discusses the role of architecture in relation to Chinese beliefs in "geomancy" (*fengshui*) – the idea that the placement of objects in space can be conducive to positive or negative energies and outcomes. A central belief with great influence in Chinese culture and practices of landscape design, geomancy provides the tenets for a shared understanding of how the landscape is both authored and read.

The construction of the Bank of China's (BOC's) new offices in Hong Kong, completed in 1990, next to the offices of HSBC, a financial competitor, triggered "a war of geomantic symbols" (2020: 439) that played out through the writing and reading of urban form. The HSBC offices, understood to occupy a particularly auspicious location in the city with a large open space in front, were now faced by the new BOC building, with its large antennae and blade-like edged design. This structure was seen by many as an aggressive, geomantic provocation. Yoon describes the tension within the following extract:

> The community started saying that the new BOC building looked like a Chinese warrior playing with a triple blade sword and two spears (antennas), and that one of the three edges was pointing at HSBC. Soon after the BOC building was completed, geomancers in Hong Kong started predicting that the BOC sword "hitting" HSBC would cause great harm to HSBC's business, and that HSBC would experience a business slump. In fact, that same year HSBC profits slid dramatically compared to previous years. HSBC was compelled to respond to this perceived geomantic "challenge" from BOC, and consulted geomancers about what geomantic actions to take. The retaliation strategy that was adopted saw the construction of two cannon shaped service winches on top of the HSBC building aiming directly at BOC building. Thus, the HSBC building was made to look like it had two cannons shooting out in defense, towards the BOC building. This particular arrangement of landscape icons engendered high tensions with serious animosity between the two bank buildings. After HSBC installed the winches

> it was reported that BOC business profits began to recede markedly. Thus the exchange of cannon balls and sword attacks was established as a geomantic battle scene, with opposing icons that were said to have damaged the business of both banks.
>
> (Yoon, 2020: 439)

The issue abated with the construction of an additional building, in 1999, between these BOC and HSBC offices, understood to mediate their "hostile relationship". Importantly, as Yoon notes, many who worked in and did business with these banks perceived the *fenshui* of the urban landscape to have a tangible effect on the companies' prospects. Thus, each "carried out geomantic iconographical landscape warfare […] in order to protect and win business from existing and future customers" (Yoon, 2020: 440).

Duncan and Duncan would also evoke literary historian Brian Stock's concept of the textual community as a means of considering the emergence of social groups organised around the adherence to, or specific readings of, particular texts (for example, different denominations of a religion constitute different textual communities). Operating via "a superstructure of agreed meaning" (Stock, 1983: 91), textual communities are characterised by a shared reading of a given text which might include "the physical and represented landscape […] read and understood" (Yoon, 2020: 432) by different groups, and in support of particular actions or values.

Thus, Duncan and Duncan demonstrate the ways in which a "landscape semiotics" drew from wider developments in literary and linguistic theory to expand a sense of landscape as text. Studies attending to open-ended, overlapping acts of interpretation went far beyond the model of signifier–signified offered by Saussure (Lindström et al, 2018) as landscapes became appreciated as assembled collections of signs operating both as direct stand-ins for the things they appeared to and abstract ideas or larger entities that they could only be represented in part or by virtue of conventional association. Also influential, after the writings of philosopher Jacques Derrida, was a sense of how the "web-like complexity" of texts meant that their interpretation navigated "a ceaseless play of infinitely unstable meanings" (Duncan and Duncan, 1988: 118). The meaning of something was always deferred in its resolution onto other symbols or things. The signs comprising landscapes were fixed neither in terms of interpretation nor significance. Rather, any reading – indeed, the material form – of landscape was fluid and changeable, defined by convention or context, and open to (possibly) violent contestation or upheaval (see Box 4.1).

Pause for thought: *Consider a landscape associated with cultural heritage, a political event, or a ceremonial practice (e.g. the Sycamore Gap, Uluru, or a town square). This could be a landscape you are geographically proximate to or one further away.*

Take a moment to consider:

1. Where/what is your landscape?
2. How can this landscape be read?
3. What sort of texts comprise or relate to this landscape?
4. How might they be considered competing/overlapping?
5. Are there any textual communities associated with the different "readings" of this landscape?

Textual Orientations

In a more direct consideration, Duncan and Duncan propose examining the relationship between given landscapes and *specific* texts such as religious texts, governmental documents, or historic urban plans. Doing so aims to ask how the meaning of landscapes is produced in relation to texts proposing, or directing, individuals' interactions with surroundings (or vice versa). In this vein, scholars have advocated consideration of the productive relationship between "words" and "worlds", each remaking the other through complicated, contingent relationships (see Daya, 2019).

There are various examples that explore this productive interplay, especially concerning "official" texts like town plans or land-use policies as shaping the future of urban areas or naturalising (making seem normal, or common sense) ideas about the city landscape, its appearance, function, and future (see Schein, 2009; Kenny, 1992). Consequently, the landscape can be understood as "discourse materialised" (Schein, 1997: 666). We can also consider texts that inform specific kinds of bodily interactions with the landscape; choreographing vision or even touch in the way our limbs and muscles encounter the material environment. Guidebooks, for instance, can be understood after theorist Sara Ahmed (2006 – see also Chapter 9) as "orientation devices", foregrounding certain features of our surroundings within perception (encouraging us to notice and respond to them) whilst relegating others to the background of our sensory attention. Injecting phenomenology (see Chapter 8) into discussions of the role of texts proves readily compatible with semiotic approaches that have long considered affect and sensory stimuli as important parts of meaning-making (see Lindström et al, 2018).

An evocative example can be seen in the role played by a host of textual resources, such as guides, videos, images, and websites, in the construction of rock-climbing landscapes. The modern activity of mountaineering finds its first expression in the nineteenth century, infused with ideas of wilderness and romantic exploration, as well as gender. As Nettleford and Stratford chart, following the first ascent of Mont Blanc (the highest peak in Western Europe) in 1786, mountains were "(re)presented as spaces of and for masculine achievement-oriented adventure" (1999: 134; see also della Dora, 2017). With many of the notable European peaks being summited and mapped by the early twentieth century, rock-climbing emerged a sport still yielding "firsts"; the maiden climb of a particular route or crag, potentially even via a particular kind of approach or sequence of bodily movements, replacing the un-ascended summit. Climbers sought routes that were challenging, offering a sense of jeopardy. Likewise, climbing discourse entrenched assumptions about which bodies were suited for this activity. Guidebooks identified routes (and those who had first plotted them) available to climb; first in florid prose and later by way of graphical "topos" (Figure 4.2) combining sketches or photos with annotations and labels. Rockfaces were recontextualised as textual compositions of holds, manoeuvres, or grade-ratings, directing the climber's gaze towards certain features, and creating a series of agreed standards or benchmarks for measuring ability (Rickly, 2017). As climbers sought and claimed new (variations of) routes, they were named and added to the record, further textualising the mountain landscape. As Nettleford and Stratford observe, "guidebook representations

Figure 4.2 Topo showing Climbing routes in Sobron, Alava, Euskalerria, Spain. Image by TxokoPirata. Wikimedia Commons, CC BY 3.0.

allow climbers to literally read rock-climbing landscapes as if they were static (albeit cumulative) texts" (1999: 137).

More recent work on climbing emphasises the interplay of bodies, rockfaces, and texts in mediating an encounter with landscape. Jillian Rickly (2017) explores, via interviews and ethnographic research involving North American climbing groups, how climbing texts and the activities of fellow climbers together coordinate gestures, foot and hand placement, and techniques that are translated into the production of a particular bodily experience of landscape (see Chapter 8). Moreover, the landscape is inscribed through its *physical* transformation as bolted routes, smears of chalk, or flaked or smoothed rock materialise the sport. As Rickly observes, "climbing is also a defacing", leaving traces and continually re-writing the environment (2017: 74). Climbers "work" a route over and over to better read, or learn, the rockface, holds, surfaces, and positioning. Diverse climbing communities ("scenes") coalesce around favoured locations and circulate through various *other* textual spaces, including films, magazines, and web forums. Each orients the climber towards a particular reading of landscape within "ideologies of the appropriate climbing assemblage" (Rickly, 2017: 77).

Literature and Landscape

Geographical location and historical context are central to the construction of any written text. Thus, cultural landscapes and their meanings are embroiled with works of *literature*. A literary text concerns some period and is written under particular historical conditions (material and representational or cultural). Moreover, literary works are *set*, in that they unfold within or address, somewhere (Peraldo, 2016). Recent multidisciplinary work on literature (as in the field of "literary geography") demonstrates numerous possible connections between the analysis of literature and landscape (Saunders, 2010). The ways in which texts (and authors) exhibit a "place-mindedness" reveal them to be acts of *mapping* the world, active in the creation of imagined geographies (see Tally, 2018).

Landscape studies have long made use of literary accounts. As discussed above, cultural landscapes were considered "readable" in part via such written works. Later, humanities-inspired spatial theories advance interest in literature for what revealed about the subjective experience of different place. Elsewhere, critical interpretations of contemporary and historical written works, informed by Marxist theory or postcolonial approaches (see Chapter 6), interrogate the ways fiction constructs and circulates ideological or "othered" versions of landscape, grounded in both real encounters and fantastical desires (e.g. in science fiction – see Balfe, 2004). As Jo Sharp notes, then, such work treats literature as "part of a complex and interlinked set of texts through which discourses of identity and difference are established and communicated" (2000: 328). This is showcased in the work of James Riding (2016) who describes the ways early-twentieth-century nature writer and poet Edward Thomas animates the South Downs as imagined and engaged with by tourist-readers.

As well as examining how works of literature re-present landscapes with vested interests in mind, or the role certain landscapes play in contextualising or informing the production and reception of texts, other scholarship has explored what Sheila Hones (2011) terms the "event" of the text. This approach understands works of literature as produced, circulated, consumed, and (re-)interpreted through ongoing social and cultural relationships of writing and reading. As Angharad Saunders argues, this means "a text is not just a finished product awaiting interpretation but is equally a set of spatial practices that combine in different ways to bring the text into being" (2013: 285). An example of the various ways in which the relationship between landscape and works of literature unfolds via this way is considered in Box 4.2.

Finally, it is clear that many landscape scholars are also increasingly experimenting with what literary writing might afford *methodologically* as part of renewed interest in the potential of empirically driven "place-writing" (see Chapters 1 and 13). In turn, literature is appreciated not merely in terms of its content as relating to specific landscapes, but how "literary craft" provides the means to "bear witness" to our encounters with and knowledge of place and landscape as part of a creative research practice (see Marr et al, 2022: 574, 577).

Box 4.2 Walter Scott and Loch Katrine

A notable example of literary works informing a dominant reading of landscape can be found in tracing the influence of early-nineteenth-century Scottish writer Sir Walter Scott (1771–1832) upon the imagining and representation of his home nation within tourism promotion.

Scott's novels, influenced by ideas of the picturesque (see Chapter 2), often featured excursions by characters into scenic, wild, sublime environments of mountains, glens, and lochs. They regularly drew upon his own excursions, deploying detailed, emotive descriptions of the landscape in prose: a form of writing termed ekphrasis. His landscapes, as Celine Sabiron notes, were "arranged like a picture"; his narrators providing "a verbal painting of the Highland landscape" in their dialogue (2016: 209). The legacy of Scott's work has been profound: cementing a reading of Scotland's landscape as rugged, wild, nobly savage, and romantic in the popular tourist imagination. His texts encouraged visits north as readers sought direct experience of the places he so evocatively narrated.

This influence is especially apparent with regard to his epic poem "The Lady of the Lake" (1810) and its impact upon tourism at Loch Katrine in the Trossachs. For Nicola Watson (2012), though the novel traverses several locations, the loch becomes a synecdoche (a literary device whereby part of a larger entity serves to represent the whole) of Scott's imagined Scotland: this landscape *is* Scotland. Moreover, Scott's verse effectively models a tourist excursion into the highland environment through the journey of its central character, Fitz-James (revealed as King James I in disguise). It choreographs the tour its reader could take, signposting experiences sought out as part of a romantic, holiday sojourn.

Though appreciated as a "beauty spot" for some decades, Scott's writing, penned after his repeat travels through the region between 1790 and 1809, contributed something akin to a nine-fold rise in visitors to Loch Katrine by the end of the year of its publication in 1810. Not only did copies of the poem prompt people to come, its descriptions and imagery were circulated through various performances (on stage, or the hiring of pipers to play on the lakeshore); signage; the inclusion of excerpts in guidebooks; and even tourist infrastructure relating to plotted events. Scott's extensive footnotes and empirical scene-setting lent the poem "a tinge of historical status" (Watson, 2012: 60); enabling production of literary maps which situated its fiction within the loch's real, physical surroundings. Such practices worked to "bind poem and place together within the nineteenth-century imagination" (Watson, 2012: 56) as Scott's writing promoted a "reading" of Scotland in line with the "highland myth" (see Chapter 2).

Pause for thought: *Can you think of an example of a landscape (or landscape "type") that you perceive, or engage with, differently as a result of reading a novel or creative work in which it is represented?*

Chapter Summary

Box 4.3 Key Concepts

Chapter 4

Post-structuralism
Semiotics
Landscape as a text
Intertextuality
Literary geographies

This chapter has considered the cultural landscape as a text (Box 4.3). Considering how this metaphor has been productively applied in scholarship, it has discussed the shift from initial conceptions of landscapes as the "unwitting autobiography", inscribed as a by-product of social activity and in need of expert decoding; towards the application of conceptual approaches drawn from literary theory. In particular, the chapter examined the application of semiotics to cultural landscape interpretation, and landscapes as constellations of signs that, individually or collectively, propose a particular "meaning" for the reader.

Under the influence of post-structural theories, and more sceptical treatment of taken-for-granted acts of reading, the idea of texts as having multiple, contested, or differently asserted interpretations also informs cultural landscape studies. In this way, the landscape text becomes a site of overlapping, intermingling, variously practised, or policed readings. Indeed, there can be no one authoritative or objective account of its meaning, nor can its meaning ever be considered closed or exhausted. Within such understanding, questions of what texts can do in, or for, cultural landscape studies are posed and answered in numerous ways. This chapter has considered how texts serve to *orient* us towards particular forms of landscape interaction, as well as how engagement with literary writings offers new ways to understand the working and significance of cultural landscapes.

Suggested Seminar Activities

Activity 1: Find an example of a text (however you interpret that term) that instructs/organises/regulates the use/interpretation of a landscape in the UK. Bring a copy of this "text" to class for discussion. In groups:

- Introduce and contextualise both the text and the landscape it refers to.
- How does this text direct the conduct or activities of landscape users?
- Through this text, do we get a sense of who the landscape is "for", and who it is not for?
- How do these texts, concerning appropriate conduct when in particular landscapes, relate to or reflect particular ideas about how the landscape should appear?
- What other places, groups, or institutions are drawn into the work that this text does?

Activity 2: In groups, discuss the paper by Yoon (2020). Begin by sharing your impressions and what you understand of the article's key points. Then, consider the following questions:

- In what ways is the landscape usefully understood as a text? How (and by whom) is it "written"? How (and by whom) is it read?
- How are different ways of writing/reading the landscape bound up with questions of power or resistance to power? How do different readings emerge as a result of conflicts between landscape users?
- How do different readings of the landscape reflect different shared assumptions/values/narratives – or "discourses" – on the part of the groups interpreting them?
- How do different readings of the landscape relate to ways of acting on, being in, or "inhabiting" the landscape?

Recommended Reading

Rickly, J. 2017. The (re)production of climbing space: Bodies, gestures, texts. *Cultural Geographies* 24(1): 69–88.

Yoon, H.-K. 2020. Iconographical landscape warfare. *Landscape Research* 45(4): 428–443.

References

Ahmed, S. 2006. *Queer Phenomenology: Orientations, Objects, Others*. Durham, NC: Duke University Press.

Balfe, M. 2004. Incredible geographies? Orientalism and genre fantasy. *Social & Cultural Geography* 5(1): 75–90.

Barthes, R. 1981. *Mythologies*. London: Granada.

Bassett, K. 2023. TripAdvisor as a 'geo-pastoral technology'. *Tourism Geographies*. Published online 30 Oct 2023. https://doi.org/10.1080/14616688.2023.2275734

Cosgrove, D. and Jackson, P. 1987. New directions in cultural geography. *Area* 19(2): 95–101.

Daniels, S. and Cosgrove, D. 1993. Spectacle and text: Landscape metaphors in cultural geography. In: Duncan, J. and Ley, D. Eds. *Place/Culture/Representation*. London: Routledge, pp. 57–77.

Daya, S. 2019. Words and worlds: Textual representation and new materialism. *Cultural Geographies* 26(3): 361–377.

Della Dora, V. 2017. *Mountain*. London: Reaktion.

Doel, M. 2023. Textual analysis. In: Clifford, N., Cope, M., Gillespie, T., and French, S. Eds. *Key Methods in Geography*. 4th Ed. London: Sage, pp. 245–261.

Duncan, J. 1990. *The City as Text: The Politics of Landscape Interpretation in the Kandyan Kingdom*. Cambridge: Cambridge University Press.

Duncan, J. 1998. Dis-orientation: On the shock of the familiar in a far-away place. In: Duncan, J. and Gregory, D. Eds. *Writes of Passage: Reading Travel Writing*. London: Routledge, pp. 159–171.

Duncan, J. and Duncan, N. 1988. (Re)reading the landscape. *Environment and Planning D: Society and space* 6(2): 117–126.

Duncan, J. and Duncan, N. 1992. Ideology and bliss: Roland Barthes and the secret histories of landscape. In: Barnes, T. and Duncan, J. Eds. *Writing Worlds: Discourse, Text and Metaphor in the Representation of Landscape*. London: Routledge, pp. 18–37.

Duncan, N. and Duncan, J. 2010. Doing landscape interpretation. In: DeLyser, D., Herbert, S., Aitken, S., Crang, M., and McDowell, L. Eds. *The SAGE Handbook of Qualitative Geography*. London: Sage, pp. 225–247.

Hones, S. 2011. Literary geography: The novel as a spatial event. In: Daniels, S., DeLyser, D., Etrikin, N., and Richardson, D. Eds. *Envisioning Landscapes, Making Worlds: Geography and the Humanities*. Abingdon: Routledge, pp. 279–287.

Kenny, J. 1992. Portland's Comprehensive Plan as text: The Fred Meyer case and the polities of reading. In: Barnes, T. and Duncan, J. Eds. *Writing Worlds: Discourse, Text and Metaphor in the Representation of Landscape*. London: Routledge, pp. 176–192.

Lewis, P. 1979. Axioms for reading the landscape: Some guides to the American Scene. In: Meinig, D. Ed. *The Interpretation of Ordinary Landscapes*. Oxford: OUP, pp. 11–32.

Lindström, K., Palang, H., and Kull, K. 2018. Landscape semiotics. In: Howard, P., Thompson, I., Waterton, E., and Atha, M. Eds. *The Routledge Companion to Landscape Studies*. Abingdon: Routledge, pp. 72–90.

Marr, N., Lantto, M., Larsen, M., Judith, K., Brice, S., Phoenix, J., Oliver, C., Mason, O., and Thomas, S. 2022. Sharing the field: Reflections of more-than-human field/work encounters. *GeoHumanities* 8(2): 555–585.

McGreevy, P. 1992. Reading the texts of Niagara Falls: The metaphor of death. In: Barnes, T. and Duncan, J. Eds. *Writing Worlds: Discourse, Text & Metaphor in the Representation of Landscape*. London: Routledge, pp. 50–72.

Meinig, D. Ed. 1979. *The Interpretation of Ordinary Landscapes*. Oxford: OUP.
Morin, K. 2009. Landscape: Representing and Interpreting. In: Clifford, N., Holloway, S., Rice, S., and Valentine, G. Eds. *Key Concepts in Geography*. 2nd Ed. London: Sage, pp. 286–299.
Nettleford, P. and Stratford, E. 1999. The production of climbing landscapes-as-texts. *Australian Geographical Studies* 37(2): 130–141.
Peraldo, E. 2016. Introduction: The meeting of two practices of space: Literature and geography. In: Peraldo, E. Ed. *Literature and Geography: The Writing of Space throughout History*. Newcastle: Cambridge Scholars Publishing, pp. 1–16.
Pocock, D. Ed. 1981. *Humanistic Geography and Literature: Essays on the Experience of Place*. London: Routledge.
Rickly, J. 2017. The (re)production of climbing space: Bodies, gestures, texts. *Cultural Geographies* 24(1): 69–88.
Riding, J. 2016. A geographical biography of a nature writer. *Cultural Geographies*. 23(3): 387–399.
Sabiron, C. 2016. Walter Scott and the Geographical Novel. In: Peraldo, E. Ed. *Literature and Geography: The Writing of Space throughout History*. Newcastle: Cambridge Scholars Publishing, pp. 206–221.
Sauer, C.O. 1956. The education of a geographer. *Annals of the Association of American Geographers* 46(3): 287–299.
Saunders, A. 2010. Literary geography: Reforging the connections. *Progress in Human Geography* 34(4): 436–452.
Saunders, A. 2013. The spatial event of writing: John Galsworthy and the creation of Fraternity. *Cultural Geographies* 20(3): 285–298.
Schein, R. 1997. The place of landscape: A conceptual framework for interpreting an American scene. *Annals of the Association of American Geographers* 87(4): 660–680.
Schein, R. 2009. A methodological framework for interpreting ordinary landscapes: Lexington, Kentucky's Courthouse square. *The Geographical Review* 99(3): 377–402.
Sharp, J.P. 2000. Towards a critical analysis of fictive geographies. *Area* 32(3): 327–334.
Stock, B. 1983. *The Implications of Literacy*. Princeton, NJ: Princeton University Press.
Tally, R. 2018. *Topophrenia: Place, Narrative, and the Spatial Imagination*. Bloomington, IN: Indiana University Press.
Tuan, Y.-F. 2013. *Romantic Geography: In Search of the Sublime Landscape*. Madison, WI: University of Wisconsin Press.
Watson, N. 2012. Holiday romances; or, Loch Katrine and the literary tourist. In: Brown, I. Ed. *Literary Tourism, the Trossachs, and Walter Scott*. Glasgow: Scottish Literature International, pp. 56–69.
Yoon, H.-K. 2020. Iconographical landscape warfare. *Landscape Research* 45(4): 428–443.

5 Moral Landscapes

Introduction: Stay at Home!

In 2020, the United Kingdom, like many countries around the world, implemented a nation-wide lockdown to stem the spread of Covid-19. A "stay at home" message and emergency legal restrictions on travel and recreation radically transformed how spaces and landscapes were used. One law-enforcement organisation, Derbyshire Police Force, even shared drone footage and images on Twitter/"X" of individuals engaged in activities deemed "non-essential" and in violation of lockdown (see Derbyshire Police/@DerbysPolice, 2020). Such scenes serve to highlight the way in which these landscapes, once open for recreation, were suddenly cast in relation to a new *moral* argument about such environments and their use. The presence of humans in *that* landscape (the Peak District) at *that* time (during an international pandemic) became transgressive and threatening; something to be treated with suspicion, deterred with force, and subject to negative media attention.

By summer 2020, however, the UK had relaxed its lockdown protocols and landscapes like that commented on above became, once again, sites of leisure. In July and August, many British National Parks were visited by those keen to holiday away from home. In a still-unfolding pandemic context, however, increasing visitor numbers became a matter of concern. For example, the BBC reported "issues around camping, campervans, human waste, fires, litter and verge parking" as affecting areas across the Scottish Highlands. At Loch Morlich, in the Cairngorms National Park, the roots of an ancient "granny pine" (an old, established tree which seeds new trees) were damaged by fires blamed on tourists. "Dismayed" locals called on National Park Authorities to "get a grip on the situation" (BBC News, 2020) as park managers across Britain were at pains to stress the "responsible" action of the majority amidst reported chaos. The blame for issues was put upon a "small group", derided as "dirty campers" (Strathspey Herald, 2020).

This example highlights the ways in which cultural landscapes, their representations, and people's material interactions with them are framed in relation to social *norms*. That is, questions concerning for whom, where, when, and how such landscapes *should* exist. The sense often pervades of there being a "right way" for landscapes to appear, be managed, used, or engaged with. "Land use" becomes a question of how norms are established or adhered to (Matless, 1994),

DOI: 10.4324/9781003172451-6

and *landscape* concerns how human-environmental relations "work" within the wider narrative of society, its values, ordering, and mechanisms of power. During the Covid-19 pandemic, the issue of who was allowed to be in a National Park landscape reflected *how* one was expected to behave, morally, during this time.

Indeed, National Parks were established in Britain in 1949 to provide space for leisure. There are now 15 parks across the country (ten in England, three in Wales, and two in Scotland). They were created not only to protect the land from the perceived threat of urbanisation but also to establish and maintain a set of *values* associated with the land. Through immersion in these environments, the British citizen would come to uphold certain ideals about what Britain "was", and be moulded into a useful (i.e. fit, outgoing, practical) body (Matless, 1997). Consequently, those who did not adhere to this ideal (the "dirty campers'" of lockdown) were deemed "out of place". Thus, National Parks, like all landscapes, can be understood as produced by, and *reproducing of*, a set of held assumptions about the nation, the citizen, the environment, and the proper relationship between all three. These assumptions, narratives, frameworks, or discourses involve and extend beyond given landscapes. They are created, circulated, and reinforced by various public and private actors, organisations, and institutions. Landscapes and their users are part of the performance and production of certain types of subject, or ways of acting (good or bad; right or wrong; normal or abnormal). Across multiple scales and sites, they are materially and symbolically implicated in broader "cultural movements, debates and practices" (Wylie, 2007: 95).

The cultural landscape in this chapter, then, involves the interconnection of bodies (and how they appear or what they do), environments (how they are managed or regulated), and normative values (how they circulate and are enforced). Landscape invites questions of morality, regulation, and spatially appropriate conduct. Above all, it is a technology of power, with how it appears or is used unfolding in relation to a complex interplay of physical, social, and cultural forces. The geographer and landscape scholar David Matless (1997) usefully terms "cultures of landscape" as the various discourses, practices, institutions, traditions, debates, assumptions, values, and other things besides which coalesce to produce landscapes. It is to such "cultures" that this chapter (and that which follows – Chapter 6) attends.

Chapter Outline

This chapter understands landscapes as discursive entities that are active, evolving, and incorporating not only political or economic relations, but also social and aesthetic values. It discusses the influence of a broader "moral turn" in studies of landscape, and the application of post-structuralist ideas concerning discourse, power, and governmentality. In particular, we focus on the work of geographer David Matless and the way his writing considers the relationship between forms of self-discipline, citizenship, conduct, and landscape in the context of outdoor recreation and tourism in mid-twentieth-century Britain. As individuals or their actions are moralised, so our perception and use of different environments becomes part of the performance of agreed or enforced social norms. To emphasise how discourses of landscape are both historically produced and always changing or subject to contestation, we also examine several examples to highlight how the moral narrative discourses of landscape are continually situated, shifting, and transgressed.

"Moral Landscapes"

The notion of landscapes being "moral" – bound up with ideas of what was right, proper, *or* normal – first substantively emerged in the 1990s as scholars engaged with and adapted post-structuralist cultural theory. Of particular interest was the way in which the concept of "moral landscapes" conveys a sense of their being active in working upon the subject in various ways. That is, the form, management, or associated qualities of landscapes might instil (or imply) a particular (right) way of "being" *in* them. "Landscape", as a moral entity, moves from noun to *verb* – it is not (merely) a thing, but does things.

Such an active understanding of landscape as a doing verb is reflected in the work of David Matless. Across the influential *Landscape and Englishness*, Matless traced the historic relationship between a range of landscape practices and emerging forms of national identity. He proposed attention to "cultures of landscape" as not only comprising those artefacts like artworks or paintings but associated values, narratives, and ways of being in landscape. For Matless, landscape was produced cumulatively through what was written, said, or practised about – or upon – it. A host of expressions or declarations on various topics that involved the environment, combined with material actions, served to *define* landscapes in terms of what they meant, how they were known, and their normative qualities: a *discourse* of landscape. As Wylie describes, landscape is thus named "a cultural production composed of actions, beliefs and practice" rather than "a stage on which practices are played out" (2007: 115). Its conceptualisation was inseparable from those practices which created and framed it, whether they be boating, climbing, hiking, zoning, planning, or policing. As Matless argues:

> the question of what landscape "is" or "means" can always be subsumed in the question of how it works; as a vehicle of social and self-identity, as a site for the claiming of cultural authority, as a generator of profit, as a space for different kinds of living.
>
> (1998: 12)

Therefore, attention to discourse shifted from critically dismantling representations and their ideological purpose (see Chapters 2 and 3) towards the question of how meaning was made and sustained through context-specific, practical activities of use or management. Neither purely artistic nor elite constructions put to service by a landed establishment in its efforts to naturalise a vision of society and economy, Matless' "cultures of landscape" denoted specific, lived, everyday forms of landscape activity.

For Matless, much previous landscape scholarship "missed the modernity" (1998: 16) via its focus on picturesque, pretty, painted environments. Reframing landscape as active and in use focused attention on contemporary values and tensions, whilst mindful of inherited or reworked associations from the past. For example, Oliver Dunnett's (2015) exploration of the British Campaign for Dark Skies highlights how a historical perspective on light pollution debates helps us understand the moral positions taken around the management of landscape and illumination today. Matless's approach thus emphasises how the perceptions and values attached to rural places are bound up with changes unfolding across different (e.g. urban) places. The idea of the rural was implicated in the production of a national consciousness that embroiled specific landscapes. Urban (and rural) social subjects, in turn, engaged those landscapes in the performance or expression of their identity in ways that reinforce or challenge such perceptions and values. Indeed, as discussed in Box 5.1, the moral associations attached to landscape might implicate any number of facets of one's identity.

Box 5.1 Swampification

Combining interest in Black ecologies – "a subfield that aims to illuminate conditions and relations Black people have with/in ecological and social worlds" (Vickers, 2023: 1674) – and moral geographies, Morgan P. Vickers examines how contemporary perceptions of swamp landscapes entwine historic experiences and narratives of race, environment, and oppression. Discussing the transformation of hundreds of acres of American swampland for navigation, electricity generation, and afforestation as part of the 1930s New Deal and its associated infrastructure projects, such "development", as Vickers notes, involved the "largest land clearing project in the US history" whereby "the state inundated the homelands of 901 families, more than 800 of whom were Black" (2023: 1674).

Considering the landscape's moral framing, Vickers highlights how particular imaginaries and associations were used to justify the destruction of swampland. Terming this process "swampification", they describe how "governments, corporations, and the press socially (re)invented swamplands as uninhabitable spaces of death and disease to justify their destruction" (Vickers, 2023: 1675). Cast as areas with high levels of malaria, the displacement of a predominantly Black population was subsequently justified in terms of health and the control of the disease. The imaginaries of swampland promoted by the state and developers operated on a logic whereby certain landscapes were worthy of inhabitation while others were condemned as uncivilisable, "unruly" (Vickers, 2023: 1678), or undesirable. Thus, "[t]he notion that swamplands are uninhabitable places devoid of life" is recognised as "a racial capitalist construction tied to myths of Blackness as pestilence and Black landscapes as deficient geographies, or places difficult to access, inhabit, and develop" (Vickers, 2023: 1676).

Moral judgements about swamp landscapes also entailed judgements about their inhabitants. A wider narrative concerning such swampland's (un) economic productivity bled into the argument that it was poorly managed. The "improper" use of the landscape by its predominantly Black population in turn found them wanting according to such standards, erasing any sense of how the land *was* managed and legitimating the violent removal of residents from situated, historic "ecosocial worlds" (Vickers, 2023: 1677). Ultimately, then,

> [t]o "swampify" the Lowland South through the Santee-Cooper Project was to mythologize the swamps as unruly, disorderly, and uninhabitable to civilize the ecologies and the inhabitants on the path to New Deal progress.
>
> (Vickers, 2023: 1678)

Pause for thought: *Can you think of or identify another example of where the use of a landscape by one group is recast as "misuse" by another to justify its transformation or management in some way?*

Attending to the ways in which contemporary landscape values emerge through particular historical processes underlines that landscapes are not fixed. Rather, their meaning and form are layered, and exist to be encountered as an "unstable heterogeneity" (Matless, 1998: 21). Drawing here from Michel Foucault's genealogical approach (see Box 5.2 and discussed below), Matless, and most subsequent studies of a moral landscape, thus often involve some element of historical methodology (such as archival work) to uncover and trace the formation of the norms, associated qualities, and kinds of subjects, institutions, or management approaches shaping them today.

Box 5.2 Defining a "Genealogical" Approach

Michel Foucault's writing embodied an interest in how societies were governed and organised. His commitment to historical study enabled him to demonstrate how such social formations were contingent – the outcome of context – rather than inevitable. Central to his practice was historical-archival research to excavate and trace the emergence and development of key institutions, practices, ideas, or narratives – for example, in the fields of psychiatric medicine or the development of a modern prison system – as they established certain frameworks of understanding, held truths, and senses of the norm. As he describes:

> this is what I would call genealogy, that is, a form of history which can account for the constitution of knowledges, discourses, domains of objects etc., without having to make reference to a subject which is either transcendental in relation to a field of events or runs in its empty sameness throughout the course of history.
>
> (Foucault, 1980: 117)

Foucault's intention was to follow back through history those threads that revealed how a particular, often dominant or taken for granted, assumption about social life or existence (perhaps central to the organisation of society, or its values) today presented as universal was in fact borne of or emergent from particular conditions in the past. His interest was also applied to the making of the human as a "subject" in different ways, as their body, behaviour, or sense of self was disciplined and worked upon by external forces in ways that served various interests or ways of organising the social milieu. Applied to landscape, such an approach therefore examines how particular ideas – or *discourses* – of landscape form and take hold.

Conceptualising landscapes as "moral" does not deny the role of economic and political ideology. Rather, it is to think about power differently. Linking consideration of landscape – its appearance, use, and management – with subjectivity – how people behave, are conceived of, and identify themselves – shifts the sense of power from being a force applied "from above" (and resisted below), to its operating via more diffuse forms of (self-)discipline and adherence to norms (Matless, 2000; Sisser et al, 2016). A genealogical tracing of landscape "cultures" – as collections of narratives or held ideas framing how they are perceived, understood, or acted upon – enables one to "consider the production of the bodily subject through landscape, where that landscape is not simply a backdrop to action but a culturally charged object" (Matless, 2000: 142). To do so places the application of power as it occurs in and through *specific environments*; a sense of what they "should" be like, or who they are for, and how such frameworks prove historically influential and pernicious (see Chapter 6).

Pause for thought: *Think of your own examples of how narratives of conduct are attached to specific landscapes. Where have you felt you have to act a certain way? In a playpark? A cemetery? A specifically designated landscape (e.g. a world heritage site)?*

Theorising Landscape as Discursive

Thinking about landscapes and cultural relationships in terms of discourse – something which became widespread during the 1980s and 1990s as the work of Foucault (and others) was applied across the humanities and social sciences – has important antecedents. A notion of landscape being produced in relation to ideas of social "morality" are "explicit" in the urban studies characterising of sociology's Chicago School in the early-twentieth century (Holloway and Hubbard, 2001: 200). Early in landscape studies' "cultural turn", Stephen Daniels (1982) would examine the designed gardens of Humphrey Repton and Lancelot "Capability" Brown in terms of the "morality of landscape" (a proper way of being within them) integral to how such spaces aimed to produce social harmony (see Chapter 2, see also Smith, 1997 and Setten and Brown, 2009 discussion of moral geographies). Therefore, interest in how landscapes are shaped in relation to norms runs deep. Regardless, this way of thinking landscape would develop considerably in the application of key poststructural ideas and approaches, particularly that of the analysis of discourse, as we now explore.

Discourse, Power and Governmentality

In contrast to the structuralist underpinnings of some previously examined approaches (e.g. the explanation of how landscapes are seen, read, or constituted in relation to underlying mechanisms of economy or ideology – see Chapters 2 and 3), examination of moral landscapes proceeds from the intellectual footing of *poststructuralism* (see Chapter 4). There is interest in the politics and power relations that support acts of representation, interpretation, and practice in a world with no single, settled meaning. Whilst, for a landscape scholar like Matless, "theory does not [...] have a strong explicit presence", this does not mean it is without a conceptual agenda (see Merriman, 2017: 181). Indeed, the quote from Foucault opening the introduction of *Landscape and Englishness* emphasises its genealogical treatment of "Englishness" and environment. Yet Matless describes his work as "theoretical less in terms of its evident language than in an angle of approach" (Matless, 1998: 21). That the act of attending to the specifics of how land use is framed, contested, and unfolds by way of accounts and narratives concerning a particular landscape makes clear the ways in which a sense of landscape "moves" and is woven across multiple registers, relations and activities (see Matless, 2016). Since many examples of "moral landscape" scholarship take post-structuralist concepts as the departure point for analysis, it is of value to briefly survey some key ideas that such works draw from the philosophy and writing of Foucault in particular. These are discourse, power, and governmentality.

Discourse

In *The Archaeology of Knowledge*, Foucault (1972: 90) uses *discourse* to refer to "the general domain of all statements, sometimes as an individualizable group of statements, and sometimes as regulated practice that accounts for a number of statements". By "the general domain of all statements" he thus refers to the group of statements, utterances, or expressions (textual or representations, but also practised and "embodied" – see Chapter 6) that provide a subject with the language for talking about, and knowing, a given phenomenon. In other words, "discourse" names the whole "field" of knowledge, or frame of reference, by which that topic is made sense of, treated, and engaged.

Institutions, for Foucault, are central to the reproduction, circulation, and maintenance of discourses. For example, the United Nations is a significant institution that, via its Development Programme, associated conferences (such as the Brundtland Commission), reports, speeches, and regulatory frameworks, is involved in elaborating, promoting, and enforcing a particular discourse of "sustainability". That is, what sustainability is understood to mean; as well as what it looks like; how it is measured, evaluated or practised; and – crucially – what it is not. These aspects of "sustainability" constitute its discourse and are all elaborated in a significant way through the activities or interventions of an institution like the UN. Via institutions, particular discourses take hold and become the basis for governing society, knowledge, and landscapes.

Discourses do not act on landscapes already "out there". Landscapes do not "come before" discourses; being changed or acted on by them. Rather, landscapes are defined, understood, and perceived *as landscapes* only in the context of those discourses or discursive frames that *create* or define them as "landscape". In other words, landscapes are comprehended within specific "cultures of landscape". Discourses set the bounds of things and how they are understood to manifest themselves. They prescribe some actions as normal or natural, and others as abnormal, unnatural, and "out of place".

Power

Marxist accounts of landscape, such as that offered by Don Mitchell (2000), conceive of power in negative terms as that which is used to limit, diminish, or reduce. Thus, power over (or in) landscape involves its organisation to support relations of domination and inequality by certain groups that *possess* power (the means to compel or face change upon others) over and against those who are wholly or partly "powerless". For Foucault and those influenced by his work, however, power is understood differently: as a *productive* force that "produces reality" (Foucault, 1977: 155).

What does it mean to think about power as productive? For Foucault, the exercise of power has an impact on life and the conditions for living, creating or enabling ways of being or thinking as much as denying them. For example, institutions such as the school or the prison, through acts of discipline, training, the imposition of rules and manners, and the enforcing of certain patterns for applying or evaluating knowledge, *produce* new kinds of bodies, ways of acting, and thinking – new forms of subjectivity. Moreover, they do so in aid of a state-wide agenda of productivity. The modern school aims to produce a body-subject that is useful to wider society; the prison aims to rehabilitate "abnormal" or unproductive subjects into those that are normal and productive (as opposed to the noose, which simply destroys them).

Across Foucault's work, power does not emanate from a single source. Rather, it comes from and circulates through all aspects of a social context. In Foucault's words: "power is everywhere, not because it embraces everything, but because it *comes from everywhere*" (Foucault, 1978: 93). The norms expressed in writing, laws, culture, and the design of the environment prompt us to act in a particular manner. In acting in accordance with norms, we reinforce their status *as* norms. In terms of landscape, this conceptualising of power has influenced understandings of how space and culture manifest one another and the orderings we live by (see Mitchell, 2002; Sisser et al, 2016 – and Chapter 6). Power becomes an everyday phenomenon of landscape, present in the numerous small actions, which, combined, produce a "culture".

Governmentality

Across his work on institutions, Foucault has examined the ways in which different "knowledge regimes" produce particular social "subjects". For example, the institution of Psychiatry and its discourse of "madness" produce a sense of the "mad" subject, their characteristics and behaviours. Also of interest for Foucault was the way in which institutions and their efforts to know, intervene in, or regulate individuals and social groups and environments established and reproduced norms: either as an "average" state of being, or as a sense of how things should be expected to be.

Such norms impact the creation and management of different subjects as, in the above examples, institutions including the school or prison aim to produce the subjects that "fit" with a wider society and its aims. At a more mundane level, we engage in all manner of acts of self-disciplining to adhere to norms in space, like hushing our voice in an art gallery. A sense of what is (or is not) appropriate within different surroundings – as well as the ways in which particular those surroundings might be designed or used to promote certain behaviours, and thus "improve" us as subjects more aligned with their norms – helps us appreciate how they are defined, or contextualised, within a wider discursive framework.

Governmentality, as a concept for Foucault, refers to the way in which individuals within society come to operate in line with norms, self-policing, and acting in ways that serve state interests. This is the ultimate aim of institutional engagement with individuals: subjects that self-regulate and act in accordance with discursive norms to be useful and "good". As John Wylie summarises, here Foucault's work gives us a language to consider "the conduct of the self from both without and within" (Wylie, 2007: 114) in relation to landscape, as good or bad actors (in a moral sense) become distinct in terms of those who abide by (or not) the rules of the day.

The overview given here has been, at best, brief. These are complex ideas. Nevertheless, we hope that these conceptual coordinates animate a more fulsome appreciation of what moral accounts of landscape are attempting to do in their consideration of landscapes as discursive entities. Below, we consider more practical examples of thinking with moral discourses and norms in landscape studies.

Moral Landscapes in Practice

Landscapes can be understood as historically situated, cultural entities which produce and reproduce particular discourses and subjectivities. Landscapes not only *reflect* moral (or other) values but are produced by, and reproduce, such values as they work to enact and reinforce norms of conduct in place. The following section provides examples of the ways in which landscape "rules" are established, produce bodies, and become subject to change.

Codes of Conduct

Many landscapes have associated "codes of conduct": written or (maybe more often) unwritten rules concerning what should happen within them, when, where, and how. Such codes have emerged and become taken for granted over time. Outdoor swimming, for example, as explored by Charlotte Bates and Kate Moles (2024: 891), produces landscapes in which the "proximity to nakedness" is negotiated in specific ways. Unlike in other swimming locales (the indoor pool), swimmers' bodies are not screened by changing rooms or cubicles. Consequently,

> the lack of clear zones disrupts expectations around where and how people get changed, and there are no material markers to shape interactions [..] Nakedness becomes essential and accepted, and the negative emotions we often associate with nudity [...] are washed away by the desire and necessity to get into the water, and to get warm afterwards.
>
> (Bates and Moles, 2024: 892–893)

Nakedness in the landscape is moralised differently as unacceptable, in different locations or contexts. Different sets of landscape practices – like wild swimming – carry different norms, produce different subjects, and are shaped by different institutions or organisations (for example the community-run *Wild Swimming UK* website, which maps swimming sites).

Continuing the theme of water, the Norfolk Broadlands – an area of over 300 km^2 comprising many interlinked rivers, canals, and lakes in east England – has offered Matless (1994) a rich site for examining moral landscape conduct in the explicit sense of "different ways of being in the world and the reactions of others to them" (1994: 127). Today, a National Park, designated for over 125 navigable, lock-free waterways amidst picturesque rural countryside, the framing of the Broads as a "landscape of leisure" has historically positioned certain practices as variously appropriate. Drawing on writings of key regional commentators, such as Cambridge-based professor of Geography James Alfred Steers (author of the National Survey on Coasts) and ecologist Arthur Tansley, Matless highlights how concerns over "deviant" behaviour by visitors to the area prompted calls to "educate" those seeking to visit the Broads in what were held to be "proper pleasures" (Matless, 1997: 142), such that the character of the environment would not be spoiled.

The example of discourses around tourist boat hire effectively illustrates such tensions. Boat hire on the Broads had been practised by visitors since the 1880s, yet it was primarily a middle- or upper-class pursuit until after the Second World War. From the late 1940s, with greater car ownership, disposable income, entitlement to holiday pay, and enthusiasm for the outdoors, large numbers of working-class visitors entered rural landscapes. Such "trippers" were viewed negatively; their activities were seen as "vulgar". A narrative around the proper way of boating on the Broads was steeped in "nostalgia [...] for a Broadland of the 1880's without noisy launches laden with trippers" (Matless, 1994: 133). As in Vickers' account of "Swampification" above (Box 5.1), certain groups (the urban working class) were themselves deemed "improper" by virtue of acting "improperly" within the landscape, authorising their exclusion in order to maintain propriety. Activities considered acceptable, such as sailing, were figured as pursuits by which "mentally and physically alert bodies [could] occupy the desired public space" (Matless, 1994: 134) and become further "improved".

Thus, the cultural landscape names a "culture of nature" whereby engaging the environment in different ways, for different purposes, embodies a performance of "national citizenship" (Matless, 1997: 141). The ideal landscape-citizen was produced via such "codes for conduct" that were "prohibitive" (in designating certain behaviours as abnormal) and "enabling" (in that they produced and promoted a particular version of the national subject) (Matless, 1994: 135). Criticisms of "bad actors" on the Broads carried with them the emphasis on the "physically, morally and spiritually" improving qualities of good forms of leisure (Matless, 1994: 135). Guidebooks prescribed, alongside places to visit or routes to walk, how visitors should carry themselves in rural places. Thus, as Matless argues, by translating norms of action into defined codes of conduct, subsequently circulated and promoted through tourist and institutional materials (for example the *Countryside Code*), "the Broads [as many other landscapes – see Matless 1997] became a giant open-air hostel, with certain practices forbidden in order to make a nurturing place for an improved citizen" (1994: 135).

Matless' work makes clear the link between the formation of the "citizen" and the discursive historical workings of (British) landscape. Others have explored the "moralisation of citizenship" in contemporary, non-British contexts. For example, Anderson et al (2023) have examined the discourses through which *friluftsliv* (participation in outdoor recreational activity) is encouraged in Norway, and its assumption that such practices facilitate migrant populations' assimilation into the Norwegian language and culture. As they note, a taken-for-granted link between participation and inclusion obscures the diversity of outdoor activities that take place, the relationships they produce, and the ways in which a discourse of citizenship and inclusion filters into regulation. Indeed, migrant participation in the Norwegian outdoors must confront "unwritten rules about how to behave" that "can work against, rather than towards, inclusion" (Anderson et al, 2023: 1).

Ultimately, the narrative of "*friluftsliv*" produces "a complex and contested moral landscape" (2023: 2) that sets expectations around how bodies move, locate themselves, and experience the world in ways that do not necessarily map onto the experience of all (see also Chapter 9). Moreover, being at times more visible (by virtue of an encoded or physical difference), migrant subjects can find themselves more confined by moral codes of landscape than non-immigrant Norwegians, who are able to more frequently and easily transgress them. Thus, how norms operate, where, and from whom, is a contingent and open matter.

These examples demonstrate how landscapes are constantly "shuttling between fields of reference" (Matless, 1998: 12). Neither "nature" nor "culture", physical nor social, they instead exist in the space between. Landscape – as a category, concept, and material entity – "productively migrates through regimes of value sometimes held apart" (Matless, 2003: 231). This tendency of landscape to move – elusive, with meaning deferred from thought to image to action; circulating, reproduced and reinforced – is what makes it both difficult to define and full of "analytical potential" (Matless, 1998: 12).

Moral Landscapes in Flux

Moral landscapes illustrate what Matless terms our "normative ecology of pleasures" (1997: 152): the ways in which activities are organised in relation to each other as things or activities to be enjoyed, and those which should not be. Within such divisions, there can solidify a sense of unbridgeable cultures can solidify: the idea that some people simply *do not belong* in certain landscapes. The "dirty campers" of our introduction will never not be out of place in the National Park.

Yet, the recognition of there being normal and abnormal subjects reveals a world of plurality: different ways of acting and valuing landscape that become visible and knowable in such moments of transgression. The conflicts which follow have "implications for who can stake a cultural claim over places" (Setten and Brown, 2009: 192). Box 5.3 on the geography of illumination demonstrates, drawing on the work of Tim Edensor and Steve Millington, how questions about the regulation of landscape can open onto questions about social norms more generally, and their propensity to change, endure, and be asserted or resisted by those with differing levels of authority.

Landscapes are not fixed but in flux; the matter of who has the authority to define their working, and how, is a question for empirical study. Seemingly fixed discourses of environmental conduct can be revealed as historically impermanent or contingent. Returning to our opening example, the Covid-19 lockdowns of 2020 radically rewrote the moral landscape of British National Parks (and other places besides). The good citizen became one who *did not visit*. This is an extreme example, of course, and shifts in the norms defining the production of cultural landscapes may take far longer. This does not mean, however, that norms are not consistently being subverted. Again, in Britain, we can witness ideas of who landscape is for (or not) being at times publicly challenged by the efforts of organisations such as "Black

Girls Hike" and attempts to promote inclusion and representation within outdoor recreation and challenge normative versions of landscape (see www.bghuk.com/ – also the work of Phiona Stanley discussed in Chapter 13). Beyond the UK, Oprah Winfrey's 2018 visit to Yosemite National Park in the US forced the media to acknowledge that just 2% of all National Park visitors were African American. As ranger Shelton Johnson noted, "all Americans own the National Parks, but not all Americans are here". Not-for-profit organisations like Outdoor Afro thus seek to celebrate and inspire Black connections to nature and, in turn, expand US conceptions of "the outdoors" to include Black experiences (ABC30 Action News, 16 February 2018). Such (re)presentations of the who, why, and where in respect to landscape might yet shift landscape's discourse, and its boundaries of normative inclusion.

Pause for thought: *Moral landscapes are based on sets of assumptions that are made and remade continuously by the society of which they are part. The Covid-19 lockdowns during 2020 and 2021 rapidly remade the moral landscapes of many places, albeit temporarily. Consider this abrupt change: can you think of any other examples where there has been a similarly sudden shift in the moral associations attributed to (being in) a given landscape?*

Changes in landscape discourse are not only temporal (time-bound) but also geographical (changed across space). Indeed, different norms might operate at, or shift between, different *scales*. One of this book's authors has explored this aspect of moral landscapes in work on Scottish Mountain Bothies. As she demonstrates, changes in norms and moral conventions shift between bothies and bothy-users in relation to, for example, the altitude of a bothy landscape (Hunt, 2017). The moral landscape is more amorphous than it may appear, not least as landscape discourses are subject to resistance and subversion. As Matless emphasises in *Landscape and Englishness*, cultures of landscape exist in *plural*. Understanding the ways in which moral landscapes operate is also an incitement, in the words of Setten and Brown (2009: 192), to "denaturalise and disrupt the existing order and its taken-for-granted assumptions underlying who and what is included and excluded in the use of everyday space".

Box 5.3 Diversifying Discourses: Christmas Lights

Moral landscapes can be found in urban settings as much as in rural ones. Tim Edensor and Steve Millington's (2009) discussion of festive decorations in the British suburb explores the ways in which the hanging of lights at Christmas time reveals both aesthetic norms (and their enforcement) *as well as* how such conventions are transgressed to produce a diverse, illuminated landscape. Indeed, Christmas lights (Figure 5.1) provoke much cultural

Figure 5.1 Christmas lights on a house in New Street by Steve Daniels, 2013, Bicester, UK. Image by Steven Daniels. Wikimedia Commons, CC BY-SA 2.0.

interest, with conflict instigated by "bad" lights providing the basis for representations of (sub)urban spaces across culture (e.g. the 2006 film, *Deck the Halls*) and news media (see reporting on the Whitehouse Christmas, 2017, derided as poor taste). For Edensor and Millington, outdoor Christmas lighting – particularly that festooning private residences – has become a site of class contestation, with media portrayals of extravagant lighting being characterised via the application of stereotypes (the "chav") in such a way as to mobilise the "vilification" of those transgressing the moral norms of these landscapes.

Edensor and Millington offer a discourse analysis of media representations and online social media commentary to excavate the dominant discourses mediating decoration. Discussion concerning the "respectable" Christmas landscape reveals the festive period to operate via a deeply classed set of aesthetic sensibilities. Elaborate, "tacky" displays are often associated with (stigmatised) social housing. Such distaste also expresses psychological anxieties of tasteful spaces being infected with poor taste: there is fear from some that bad lights might infiltrate their neighbourhoods. As Edensor and Millington highlight, narratives about lights are "all the time reinforcing particular cultural values that masquerade as 'common-sense'", wherein good taste requires a rejection of "less developed" impulses towards "bling" or "excessive" displays.

Investigating acts of transgression further, Edensor and Millington explore the opinions and perspectives of those whose houses are targeted with ire in the production of the festive landscape. In turn, they uncover a *counter discourse* of festivity wherein those who are critical of extravagant lighting become framed as abnormally *un*festive: "just grumpy, grumpy people" (Deborah, 30–39). Those behind the displays operate according to different aesthetic conventions ("you've got to have different colours, it's no good without a mixture of different colours" [Albert, 60–70]) associated with celebration, social connection, and community. As one interview subject, who had "a shite childhood" puts it, "I want to make sure the kids remember the Christmases as happy times" (Stuart, 30–39). The turning on of the lights becomes a moment for the street to "party".

Such debates over "how nocturnal urban space should be appropriately regulated" (Edensor and Millington, 2009: 108) reveal how moral landscapes are reproduced but are also open to contestation or transgression by alternative frameworks. This work demonstrates that researching landscape discourses means engaging with more than just dominant or majority perspectives to recognise, in this case, that "working-class culture is not point zero of culture; rather it has a different value system, one not recognised by the dominant symbolic economy" (Skeggs, 2004: 136).

Chapter Summary

Box 5.4 Key Concepts

Chapter 5

Moral landscape
Norms
Discourse
Governmentality
Power

Landscape is amorphous: it is made and remade through our actions as it also works upon us (Box 5.4). Questions of who can use landscape, for what purposes, or why are elaborated by way of an attention to moral landscapes. This chapter's focus on morality has highlighted how cultural landscapes are discursive entities, bound up with norms that intersect with particular institutional practices and serve to cultivate forms of subjectivity. It makes clear how landscape can be read as a question of "propriety" (Wylie, 2007). Shaped by ideas of "proper/improper", "right/wrong", "good/bad", or "approved of/condemned", such normative framings impact how they look, are understood, and are represented more widely.

A "lived landscape becomes a moral landscape" (Setten, 2004: 420), but it does not do so in a way that is homogenous or singular. Rather, as a product of discussion and practice, resistance and subversion, transgression and context, landscapes reveal themselves as heterogeneous, overlain with multiple "cultures" or discourses that jostle and exert themselves in different moments: landscapes are always in flux and in need of (re)asserting. The norms of landscape are determined by individuals, groups, institutions, representations, and practical acts: thus, "landscape is bounded by the people shaping it" (Setten, 2004: 390). The question of how a landscape works, then, is up for debate; changing over time, space, and in relation to the frameworks with which we seek to understand it.

Suggested Seminar Activities

Activity 1: Bring an image of a landscape to class (these could also be provided). As a class, discuss any norms, rules, or regulations that might be attributed to or associated with those landscapes, which would determine how you act within them and who they are for.

Activity 2: Moral landscapes scholarship has also emphasised the way in which those claiming authority over landscape also come to define its *soundscape.* Matless (2000)'s work on the Norfolk Broads, for example, demonstrates that the gramophone was unacceptable whilst bird song was celebrated as part of regional identity or character. By association, the particular "cultures" or social groups associated with certain sounds are "tuned out" of the region by those who seek to define it via a moralising of sonic phenomena. In this way, "the making of the broads as a nature region" is bound up with the changing "nature of Broadlands as a social space" (Matless, 2000: 144).

Task:

Split into three groups and each watch one of the clips below while in the same space (you should be able to overhear each other). While listening note down: (a) the discourses, narratives, or "norms" that each clip inscribes with regard to the sources, forms, and experiences of sound on the Norfolk Broads.

- **Video 1)** Promotional tourism video: "Norfolk Broads – Visit Norfolk, England" www.youtube.com/watch?v=ZpkdFRCSqBA
- **Video 2)** User-uploaded video: "Norfolk Broads Ranworth dawn bird song" www.youtube.com/watch?v=vGdRtEiHbr8
- **Video 3)** User-uploaded video: "Norfolk Broads 2015 Boat Party" www.youtube.com/watch?v=Gi21qNAkEh4

Discuss:

1. What were the findings of each group? How do they differ?
2. What was the impact of listening concurrently?
3. These videos demonstrate that questions of propriety around noise exist today in the Norfolk Broads as much as they did in the mid-twentieth century, reinforcing the same exclusions as "efforts to make Broadland a popular nature space downplay another form of popular culture" (Matless, 2000: 143). Can you think of any other landscapes where certain sounds are welcomed, and others vilified?

Recommended Reading

Matless, D. 2016. *Landscape and Englishness*. London: Reaktion Books.

Sisser, J.M., Nelson, K.C., Larson, K.L., Ogden, L.A., Polsky, C., and Chowdhury, R.R. 2016. Lawn enforcement: How municipal policies and neighborhood norms influence homeowner residential landscape management. *Landscape and Urban Planning* 150: 16–25.

Vickers, M. 2023. On swampification: Black ecologies, moral geographies, and racialized swampland destruction. *Annals of the American Association of Geographers* 113(7): 1674–1681.

References

Anderson, S. 2023. 'That's not how you do it': Moralising the migrant in the Norwegian outdoors. *Geoforum* 147.

Bates, C. and Moles, K. 2024. Bobbing in the park: Wild swimming, conviviality and belonging. *Leisure Studies* 43: 887–899.

BBC News. 7 September 2020. Fires and human waste left after disco in Cairngorms. www.bbc.co.uk/news/uk-scotland-highlands-islands-54058545. Accessed on 27 August 2024.

Daniels, S. 1982. Humphrey repton and the morality of landscape. In: Gold, J. and Burgess, J. Eds. *Valued Environments*. London: Allen and Unwin, pp. 124–125.

Daniels, S. 2011. Great balls of fire – Envisioning the brilliant meteor of 1783. In: Daniels, S., DeLyser, D., Entrikin, D., and Richardson, D. Eds. *Envisioning Landscapes, Making Worlds – Geography and the Humanities*. London: Routledge, pp. 155–169.

Derbyshire Police/@DerbysPolice. 26 March 2020. Despite posts yesterday highlighting issues of people still visiting the #PeakDistrict despite government guidance, the message is still not getting through [...]. Twitter/X. https://x.com/DerbysPolice/status/1243168931503882241. Accessed on 4 July 2025.

Dunnett, O. 2015. Contested landscapes: The moral geographies of light pollution in Britain. *Cultural Geographies* 22: 619–636.

Edensor, T. and Millington, S. 2009. Illuminations, class identities and the contested landscapes of Christmas. *Sociology* 43: 103–121.

Foucault, M. 1972. *The Archaeology of Knowledge & the Discourse on Language*. New York: Pantheon.

Foucault, M. 1977. *The Order of Things*. London: Routledge.

Foucault, M. 1979. *The History of Sexuality*, vol. 1. London: Allen Lane.

Foucault, M. 1980. *Power/Knowledge: Selected Interviews and Other Writings 1972–1977*, edited by Colin Gordon. New York: Pantheon Books.

Holden, J. 2015. *Norfolk broads Ranworth dawn bird song*. Available at: www.youtube.com/watch?v=vGdRtEiHbr8. Accessed 15th January 2024.

Holloway, L. and Hubbard, P. 2001. *People and Place: The Extraordinary Geographies of Everyday Life*. Essex: Pearson Education Limited.

Hunt, R. 2017. Books, bothies and thinking in place: A new contribution to geographies of the book. *Cultural Geographies* 24: 513–538.

Matless, D. 1994. Moral geography in broadland. *Ecumene* 1: 127–155.

Matless, D. 1997. Moral geographies of English landscape. *Landscape Research* 22: 141–155.

Matless, D. 1998. *Landscape and Englishness*. London: Reaktion Books.

Matless, D. 2000. Action and noise over a hundred years: The making of a nature region. *Body and Society* 6: 141–165.

Matless, D. 2003. Introduction: The properties of landscape. In: Anderson, K., Domosh, M., Pile, S. and Thrift, N. Eds. *Handbook of Cultural Geography*. London: Sage, pp. 227–232.

Matthews, J. 2015. *Norfolk broads 2015 boat party*. Available at: www.youtube.com/watch?v=Gi21qNAkEh4. Accessed 15 January 2024.

Merriman, P. 2017 'Landscape and Englishness' ... and me. *Cultural Geographies* 24: 181–184.

Mitchell, D. 2000. *Cultural Geography: A Critical Introduction*. Oxford: Basil Blackwell.

Mitchell. W.J.T. Ed. 2002. *Landscape and Power*. 2nd Ed. Chicago, IL: University of Chicago Press.

Rebanks, J. 2015. *The Shepherds Life*. Penguin, UK: Allen Lane.

Setten, G. 2004. The habitus, the rule and the moral landscape. *Cultural Geographies* 11: 389–415.

Setten, G. and Brown, K. 2009. Moral landscapes. In: Kitchen, R. and Thrift, N. Eds. *International Encyclopaedia of Human Geography*. Oxford: Elsevier Science, pp. 191–195.

Sisser, J.M., Nelson, K.C., Larson, K.L., Ogden, L.A., Polsky, C., and Chowdhury, R.R. 2016. Lawn enforcement: How municipal policies and neighborhood norms influence homeowner residential landscape management. *Landscape and Urban Planning* 150: 16–25.

Smith, D. 1997. Geography and ethics: A moral turn? *Progress in Human Geography* 21: 583–590.

Strathspey Herald. 20 July 2020. '"Dirty Campers" at Loch Morlich spark call-outs by police and fire service'. www.strathspey-herald.co.uk/news/dirty-campers-at-loch-morlich-spark-call-outs-by-police-and-fire-service-206513/

Tolia-Kelly, D.P. 2016. *Landscape, Race and Memory: Material Ecologies of Citizenship*. Abingdon: Routledge.

Vickers, M. 2023. On swampification: Black ecologies, moral geographies, and racialized swampland destruction. *Annals of the American Association of Geographers* 113(7): 1674–1681.

VisitNorfolk. 2015. *Norfolk broads – Visit Norfolk, England*. Available at: www.youtube.com/watch?v=ZpkdFRCSqBA. Accessed 15 January 2024.

Wylie, J. 2007. *Landscape*. Oxon: Routledge.

6 Landscape and Empire

Introduction: The Complexity of Empire's Landscapes

Situated on the eastern coast of Mexico, the port town of Veracruz and its surrounding region are today characterised by an extensive grazing landscape of fields and farm animals. In the sixteenth century, however, this area was the "beach head" of Spanish colonisation as settlers arrived looking to exploit the resources and promise of the "New World". As geographer Andrew Sluyter recounts the Spanish explorers, including Hernando Cortez in 1519, described encounters with pristine, tropical lowland wilderness, prime for adaptation and rationalisation. By the end of the century, dispatches home portrayed a landscape of pasture, grazed by tens of thousands of imported animals. Around three centuries later, in the mid-1800s, the despoiled state of the Veracruz lowlands was bemoaned by observers finding not a lush, fertile landscape (as previously depicted) but one barren, thorny, and overgrazed as a consequence of alleged mismanagement (see Sluyter, 1999).

Sluyter's historical account of Veracruz's material transformation under Spanish colonisation illuminates the relationship between imperialism and landscape. Settlers conceptually (re)framed the environment as a pristine wilderness – a *tabula rasa*, or blank slate – barely touched, if at all, by the indigenous societies of the Aztec and Maya peoples living amidst it. Such a narrative was inspired by, and reinforced a sense of, lands beyond Europe as uncivilised, backward, savage places, requiring European intervention and management. This way of thinking came to be expressed through many acts of reconstruction and representation central to the project of "Empire". Thus, as Sluyter emphasises, understanding the "material-conceptual, positive feedback process" (1999: 378) by which environments distant from Europe were cast as pristine or uncivilised; the activities of colonists figured as "rationalising" or modernising; and landscapes transformed in line with colonial interests in various detrimental and unsustainable ways, is of fundamental importance to landscape studies. This is not least because the legacies of these processes continue to affect the world in the wake of Empire today.

Furthermore, as Sluyter emphasises, close examination of historical landscapes in places like Veracruz helps us understand the complexities, power relations, and multiple dimensions of colonialism and imperialism as they operated through a range of different settings. Attention to the fraught, lived histories of landscapes

DOI: 10.4324/9781003172451-7

such those as in the Veracruz lowlands, reveals their often *negotiated* character. Indeed, Sluyter and Chris Duvall's (2016) subsequent research has revealed how these environments were *already* a product of management, through burning regimes, long before the Spanish arrived, and thereafter despite laws in "New Spain" designed to curtail such practices. There is also evidence that burning practices adapted to and incorporated the knowledge of pasture management and burning held by West African enslaved peoples from areas including Senegambia, brought to the region by the Spanish. Thus, attention to landscape change can reveal an intermingling of agencies as indigenous peoples, enslaved peoples, and European colonisers all play a role in producing an environment and its imaginary.

This chapter therefore emphasises the fundamental interconnection between landscape and the practices, imaginaries, and legacies of Empire; legacies which are so significant that they directly inform many of the challenges faced by states living with the legacy of colonial rule (such as land degradation, food security, and poverty). This chapter takes the lead from the complex material-representational character of colonial landscapes, illuminated by the Veracruz example above, to consider the importance of landscape as "a key unit of analysis" (Sluyter, 2001: 410) and a conceptual-empirical focus in studies of Empire and imperialism. As becomes clear, colonial landscapes reveal themselves to be *manifolds*: complicated meeting points wherein European ideas about difference, and associated material interventions, met (often violently) with the societies and agencies of indigenous peoples to produce new spaces, knowledges, power relations, and ways of experiencing the landscape.

Chapter Outline

To address the fundamental interconnection between landscape and Empire, the chapter begins by defining the key terms Empire, imperialism, and colonialism, and their relevance for landscape studies. We then consider several ways in which the landscape has been materially and conceptually shaped by imperialism. In particular, the concept of Orientalism is examined as framing encounters with non-European "otherness". The plantation as a particular form of imperial landscape is also discussed. Finally, the chapter reflects on the implications of imperialism for landscapes in the metropole (i.e. in the imperial centre) with an examination of the "imperial city" of London. The subsequent chapter (Chapter 7) builds on all these concerns to adopt a decolonising approach to thinking about landscape beyond European colonial traditions of thought.

Reckoning with Empire

Contemporary interest in considering life in the wake of colonisation and Empire (postcolonialism), alongside the sense of needing to reckon with the pervasive material and cultural inheritance of imperial relations, activities, and knowledge (decolonisation), has resulted in a great deal of critical scrutiny regarding what

Mary Louise Pratt names "the meaning-making powers of empire" (2008: 3). Before turning the relationship between Empire, imperialism, colonialism and landscape, it is worth examining what we mean by these terms, featuring as they do throughout this chapter and associated scholarship.

Defining Empire, Imperialism, and Colonialism

International Relations scholar Alejandro Colás describes how the word "empire" derives from the Latin *imperium* (authority, command, rulership, or power); initially referring to specific political powers accorded to ruling figures in Roman society. With the expansion of Roman territory, the term came to evoke the collected peoples under Roman control. Thus, through acts of military conquest, this "abstract noun" concerning political authority transformed to name a "specific socio-political and geographical entity" (e.g. "the Roman Empire") (Colás, 2007: 5–6). Consequently, imperialism, Colás argues, constituted "the active policy and process of empire-building" (2007: 9). In turn, the settling, absorbing, and governing of distant places ("colonies") with a majority indigenous population(s) by a smaller population from the conquering society comprises the practice known as colonialism (see Phillips, 2014).

For Colás, imperialism is defined by three interlinked processes. Firstly, the *expansion of territory* by one group into that of another (including establishing of colonies) such that the former conquers and exerts control over the latter "through a combination of coercive, legal, cultural and economic mechanisms", resulting in "a larger socio-economic and political entity – an empire" (Colás, 2007: 6). Secondly, imperialism is characterised by "hierarchical rule over a periphery from a metropolitan centre" as an imperial power seeks to rule over distant peoples decreed as of lesser status from the capital (or "motherland"). As such, an empire will "spatialize power along geographical lines" (Colás, 2007: 7). Finally, empires aim to impose some form of "order", defined and exported from the ruling centre, to "civilise" and improve the periphery. They are thus always a matter of "subjection" or the cultivation of new, improved "subjects" (and appropriate environments) reflecting imperial values. Since all three processes meet resistance, imperialism "always involves the use of force" (Colás, 2007: 9) against the peoples, societies, and places it engages.

Empires have existed throughout history; emerging, expanding, and subsequently crumbling away in different geographical, political, economic, and cultural contexts. The ancient societies of the Greeks, Romans (from which the word comes), Chinese, Mongols, and others all, at various times, ruled over vast swathes of territory and their diverse inhabitants, imposing new material environments and culture. Much contemporary scholarship concern Empire and imperialism, however, tends to focus upon the imperial exploits of European nations – especially the British Empire – between the latter fifteenth and latter twentieth centuries, as well as extending into the present via a range of geopolitical practices and relations that construct, in various forms, a "colonial present" (Gregory, 2004) and forms of "neocolonialism".

Theorising Imperial Landscapes

The landscape and activities of Empire are fundamentally interlinked. Thus, landscape provides the scholar of imperialism with "a key unit of geographic analysis" for apprehending transformations of the environment in its physical and conceptual dimensions (Sluyter, 2001: 410) and their legacies. Appreciating the inertia of landscape – its capacity to make solid, and thus imbue a contingent political organisation of the environment with a sense of permanence, or taken-for-grantedness, such that it becomes the backdrop for subsequent action (see Chapter 3) – necessitates engagement with landscape's imperial histories in places like Veracruz, above. Moreover, the work of Sluyter emphasises the importance of understanding the "equally real representations" (1999: 396) of landscape, and their legacies. Such work demonstrates that the study of colonial landscapes should proceed through a relational framing that recognises the interlinked, interacting agencies of indigenous societies and groups, colonising forces (European or otherwise), and the landscape itself (as both a material context characterised by a range of processes, resources, and forces and set of ideas or aesthetic traditions) (see Bluwstein, 2021).

The following sections elaborate such connections in terms of understanding imperial landscapes via an attention to their material reorganisation, their representation in travel writing and other cultural activities and the relationship between the "dreamwork" of imperial discourses (such as Orientalism) and the landscape. Finally, attention is paid to an iconic (and notorious) landscape of colonialism: the plantation.

Material Transformations

Landscapes were materially transformed by imperialism and settler colonialism. At its most basic, the mass movement of people and goods (including animals and plants) characteristic of the Columbian Exchange (migration to the "New World" from Europe following the voyage of Christopher Columbus) led to the introduction of a range of species with long-lasting transformation of environments on both sides of the Atlantic. The material attractions of the "NeoEuropes" – Australia, the Americas – where temperate and attractive climatic, vegetational and geological conditions resulted in landscapes attractive as potential farmland (see Crosby, 2015), saw these places prized by colonising European powers, and the introduction of numerous European species as part of efforts to remake them in Europe's image. Thus, in many ways, imperialism was an *ecological* process (see Box 6.1).

Elsewhere (as discussed below), environments in the region termed the "tropics" were cast as dangerous, disease-ridden, and in need of substantial engineering in order that they might prove productive, therapeutic, and conducive to the application of colonial power (Kenny, 1995; Arnold, 2000; Baker, 2014). Supposedly toxic landscapes would be drained or cleared (including indigenous settlements) as a means of preventing outbreaks of illness. Thus, we can appreciate that understanding the relationships between landscape and Empire requires grappling with relations of material transformation as much as cultural representation or power.

Box 6.1 Ecological Imperialism

It is imperative to appreciate the long-lasting, ecological consequences of empire-building. The result of bringing species from Europe to the Americas, Australia, India, and other places – effectively reuniting biota from continents separated for hundreds of millions of years – has marked and lasting consequences. The European colonial project is in many ways entangled with processes of biology and ecology, as much as those of military force and the conceptual (re-)ordering of the world. The ability of invasive species to take hold and flourish in new lands, the role of introduced diseases in decimating indigenous populations, and the hospitability of landscapes overseas when it came to hosting European agricultural activities. Together, such phenomena lead to an appreciation of what Alfred Crosby termed "ecological imperialism" – the process by which the biota of Europe colonised the landscapes of the "New World" and other places through military and cultural imperialism:

> Forests were razed for timber and fuel and to make way for new enterprises; burgeoning herds of Old World animals grazed and overgrazed the grasslands and invaded the woodlands; and the cultivated fields of the declining Amerindian populations reverted to nature, a nature whose most aggressive plants were now exotic immigrants.
>
> (Crosby, 2015: 151)

For example, the Spanish introduced many species that dominate as weeds within Mexico's landscape today, as well as a range of "Old World" germs – particularly smallpox – that destroyed the indigenous Amerindian population. These introductions, alongside the arrival of European settlers and enslaved peoples from Western Africa, created the conditions for contemporary society in the Americas, Australia, and New Zealand. As Crosby argues about arriving European settlers to America: "It was their germs [...] that were chiefly responsible for sweeping aside the indigenes" (2015: 196). For example, the 10–14-day incubation time for smallpox meant that those appearing healthy who fled the sick and suffering often carried the disease with them, allowing its spread beyond the European "frontier" and into communities yet to encounter colonising activity. This transfer of disease was "wondrously one-sided" with no evidence to suggest that pathogens from the colonies (except venereal syphilis) arrived to ravage the populations of Europe in the same way (Cosby, 2015: 215).

> ***Pause for thought:*** *Can you identify any other examples of species introduced to Europe, or exported from Europe to other places, in the course of colonial activity? What impacts have these introductions had on their "host" ecologies?*

Writing and Representing Landscape under Empire

The definition and understanding of landscape is inextricably tied to the activities and politics of Imperialism as a political and knowledge project (Mitchell, 2002). Such a claim is readily evidenced through the substantial discussion of travel writing in the nineteenth century, often produced alongside or as part of natural-history-based excursions that sought to collate and quantify the extent of imperial territory and its varied human and non-human inhabitants. Literature scholars such as Mary-Louise Pratt (2008); geographers such as Kirsten Greer (2020); and historians such as David Arnold (2006) have variously examined the role of public-facing texts by explorers, travelling scientists, soldiers, and early colonial tourists as representing, disseminating, and reproducing ideas about non-European places and peoples. Imperialism facilitated (for some) a great deal of mobility. With travel to or from the imperial centre and various colonies in the course of enacting military governance or pursuing expeditionary activities on behalf of various organisations such as the Royal Society, Royal Geographical Society, or the Africa Association in the UK, established in the late eighteenth century.

The resulting travellers' accounts served to reproduce various ideas about colonial landscapes. On the one hand, as Greer's (2020) work on military ornithology demonstrates, colonial agents sought to collect, describe, and classify the ecologies of places visited. As she describes, solider-naturalists' efforts to obtain and circulate specimens of rare birds extended via personal and professional scientific networks to furnish knowledge of new or unfamiliar species and environments. Other travellers, as Pratt (2008) examines, produced detailed accounts of landforms, vegetation and resources, akin the kinds of "textual surveys" that Spanish officials undertook upon their arrival in Mexico, to apportion up "colonial territory" in relation to commercial and property interests in a way that brought an Empire into being for the purposes of governance within the metropolitan centre (see Hunter and Sluyter, 2011).

Equally, much travel writing, from the latter eighteenth century, was much more sentimental, subjective, and concerned with telling tales of individual heroism and adventure. Pratt describes the travels of Scottish explorer Mungo Park (Figure 6.1), a member of the Africa Association. Formed in 1788 in London and consisting of an "alliance of aristocrats and wealthy businessmen", this organisation aimed to promote further exploration of the interior of the African continent for commercial gain. In 1797, Park returned from his more than a year-long excursion in the Niger basin attempting to map the river as a potential trade route and published his account of his travels. The book sold out its first print run in a month, such was

Figure 6.1 Mungo Park sets out from Pisania, from the 1892 book *The Story of Africa and Its Explorers*. Image by Robert Brown. Wikimedia, CC0 1.0/Public Domain.

the fervour and excitement to read of his adventures and discoveries. This text, as Pratt describes, is notable for signalling "the eruption of the sentimental mode" (2008: 72) as the stereotypes of "savage" non-European peoples propagated during the slave trade gave way to a more curious and lustful gaze in the wake of abolition. The interior of the African continent became the "dark heart" that contains treasure, knowledge, or wondrous civilisations, yet undiscovered. Rather than concern itself with detailed descriptions of scenery, vegetation, or, indeed, landscape, Park's text presented him as a hero, and colonial environments as the stage for European adventure. Thus, the production of travel narratives shaped perceptions of Empire at home by casting readers in Europe as participants in an exciting project of expansion, conquest, and power as imperial subjects (Pratt, 2008).

Discourse, Landscape, Imperialism: The "Dreamwork" of Empire

The concept of discourse (see Chapters 4 and 5) offers a means of exploring how, under Empire, landscape worked to reproduce (or impose) certain ways of understanding, valuing, and perceiving the world as it was materially reorganised and lived through. For W.J.T. Mitchell (2002), landscape names a domain of power, conjuring not merely genres of representation (e.g. "ways of seeing"– see Chapter 2) but ways of *thinking*. In other words, landscape captures not just how we visually appraise the environment, but how such environments are made sense of, managed, and regulated in relation to ingrained assumptions about the world and its workings. Thus, Mitchell argues, landscape was implicated in the definition, circulation, and interpretation of numerous concepts, including civilisation, savagery, nature, culture, the urban, the rural, the developed, and the underdeveloped, central to an Imperial worldview. Indeed, specific landscapes work to promote and give form to such ideas.

Considered as a discursive entity, bound up with the question of norms, institutions, and conceptions of truth, Mitchell proposes that, under Empire, landscape's "dark side" (Mitchell, 2002: 6) is made plain as it is implicated in forms of violence, oppression, and inequality characterising the geography and history of imperialism. As he proposes:

> At a minimum we need to explore the possibility that the representation of landscape is not only a matter of internal politics and national or class ideology but also an international, global phenomenon, intimately bound up with discourses of imperialism.
>
> (Mitchell, 2002: 7)

The landscape becomes a key ideological battleground for defining and executing the parameters of an imperial perspective. For Mitchell, landscape should thus be understood as central to "the 'dreamwork' of imperialism"; its investigation offering "provocation to an enquiry" (2002: 10) regarding its legacy. Landscape scholars should therefore examine how ideas and manifestations of landscape's form, function, and meaning were propagated by colonisers, imposed onto newly encountered environments, and served to reflect back ideas of both "utopian fantasies of the perfected imperial prospect and fractured images of unresolved ambivalence and unsuppressed resistance" (2002: 10). Mitchell invites an examination of how imperialism, as a set of discourses about difference, work on and through landscapes and how, in turn, landscapes reproduced, or perhaps reworked, the relationships of imperialism. Thus, as Denis Cosgrove (2003: 264) summarises, conceptualising landscape as Empire's "dreamwork" helps grasp its existence as both an "idea" *and* a material domain inextricable from "the spatial and social perceptions, assumptions and practices that accompanied European expansion into non-European regions of the globe".

For geographer Jo Sharp (2022), investigating the landscapes that hosted activities of imperialism and colonialism reveals much about how colonising nations constructed an understanding of both themselves and those they sought to subjugate. In re-engineering non-Western environments, European powers sought to produce ordered landscapes that provided, in turn, "a justification of the colonising mission" (Cosgrove, 2003: 264). Importantly, however, such landscapes were never purely the outcome of *colonisers'* ideas and activities. Rather, characterised by negotiation and hybridity, they represented what Pratt (2008) terms "contact zones": spaces of mixing, exchange, and the emergence of new forms of culture and practice as different peoples met, interacted, and lived together. The outcomes of such interaction were felt overseas *and* at home, where new ideas, cultural activities, commodities, and landscapes also took hold.

As discussed in Chapter 4, geographer James Duncan, whose work has examined colonial landscapes in Sri Lanka, argues that discourses – as constellations of values, assumptions, and narratives comprising a way of understanding the world and one's place within it – have "an attendant landscape model" (1990: 12). Fundamental to the history of Empire, during the nineteenth century, were specific ideas about race, expressed through biological categorising, the application of theories such as evolution, medicine, and representations of "the tropics" as sites of degeneracy and disease (Arnold, 1993). It is therefore significant to explore how ways of thinking about difference inform the production of landscapes. We do so below with reference to the influential concept of Orientalism.

Orientalism and Imagined Geographies of Landscape

When travelling somewhere new we never, in fact, arrive into the unknown. Rather, before departing, and after we get there, experiences of apparently-novel surroundings are *mediated* by existing representations, associations, ideas of normality, and the knowledge carried with us and which circulate via media including guidebooks, literature, art, television, film, and academic work. Together, these mediating forces construct certain imagined geographies about the world and its contents.

This term, imagined or imaginative geography, was influentially deployed in the work of Palestinian-American literary critic Edward Said. In his 1978 treatise, *Orientalism*, Said examined Euro-American perceptions of places beyond Europe (specifically the Middle East). And argued that their grouping under the label of the "Orient" served to reveal and reflect their framing for European subjects as sites of fundamental difference or "otherness". Such a framing lay at the core of the self-definition and construction of ideas of "Westernness" (and associated concepts like "civilisation"). For Said, "from earliest times in Europe the Orient was something more than what was empirically known about it" (1978: 56). It was a geography given form in the minds, writings, practices, artworks, and institutions of Europeans in ways that "[dramatised] the distance and difference of other cultures" (Said, 1978: 59).

European societies had long depicted those living beyond their lands as different or deformed from an ideal of the (Western) human subject. Sharp (2022: 14) describes how, during the Middle Ages (and enduring thereafter – see Safier, 2014), theories inherited from the Greeks (principally Aristotle) of the Earth's division into a series of climatic zones, alongside tales of mythical beasts encountered in exotic lands, informed representations of non-European "imagined others" as variously exhibiting physiological differences, or engaged in alien activities, that inverted or challenged social norms. These "monstrous" beings blended actual empirical observations (people with darker skin; practices unfamiliar to Europeans, such as lip-stretching or yoga) with superstition, fantasy, and religious moralising. In this way, a figure of the (Oriental) other took shape as that against which European civilisation was defined. As Pratt observes: "empires create in the imperial centre of power an obsessive need to present and re-present its peripheries and its others continually to itself". In this way, imperial society "becomes dependent on its others to know itself" (Pratt, 2008: 4).

> ***Pause for thought:*** *Can you think of a contemporary example of when an environment or a cultural group has been depicted in ways which could be defined as "monstrous" to uphold certain ideals or norms about where you live?*

For Said, "Orientalism" named the pervasive and persistent discourse of imagined, hierarchical differences separating "the West" (i.e. white, European society) from places beyond. This worldview was constructed and maintained by all manner of activities, institutions, and representations, including: academic study of non-European peoples and cultures, art and literature, science, popular culture, journalism, and travel writing. Thus, "the Orient" was *spoken for* by a range of Europeans (such as doctors, writers, explorers, traders, military officers and politicians) in ways that, across various, intertextual materials, elaborated as taken-for-granted fact "a basic geographical distinction (the world is made up of two unequal halves, Orient and Occident)" (Said, 1978: 12). Thus, Orientalism was "above all, a discourse [...] produced and [existing] in an uneven exchange with various kinds of power", including political power, scholarly and scientific intellectualism, cultural values and moral codes. These objects, institutions, texts, and activities associated with reproducing this discourse served as "the lenses through which the Orient is experienced" (1978: 58).

The effect of Orientalist discourse was to draw "[a] line [...] between two continents", leaving Europe "powerful and articulate" whilst "Asia is defeated and distant" (Said, 1978: 57). Moreover, the framework of Orientalism subsumed and effaced vast differences *between* the cultures of peoples in Africa, Asia and elsewhere as they became, all, citizens of "the Orient". As the postcolonial scholar and psychiatrist Franz Fanon writes, regarding the psychological and political consequences of this discourse:

> The settler makes history [...] He is the absolute beginning: "This land was created by us"; he is the unceasing cause: "If we leave, all is lost, and the country will go back to the Middle Ages." Over against him torpid creatures, wasted by fevers, obsessed by ancestral customs, form an almost inorganic background for the innovating dynamism of colonial mercantilism.
>
> (1965: 39–40)

In a framework of absolute, hierarchical differences, "native" society (and its landscapes) were condemned to be without history prior to the arrival of colonising powers. The environments of Europe were interpreted positively as the basis for a host of aesthetic value judgements and knowledge claims regarding "good" landscapes conducive to health, civilisation, and productivity. The surroundings encountered overseas were cast as in need of *improvement*, considered under-developed, dangerous, or associated with negative traits and subjects. In this sense, as Phillips (2014: 498) argues, "imaginative geographies of empire did not just represent the [world], they helped construct it". In Box 6.2 we elaborate examples drawn from scholarship concerning British imperialism in India to demonstrate how ideas of fundamental difference informed the construction, interpretation, and consumption of urban landscapes.

Box 6.2 Urban Landscapes in Imperial India

The British Empire in India: Beginning in earnest in the sixteenth century, accelerating as expeditions and exploration activities secured trade routes, resources, and territory during the seventeenth and eighteenth centuries, at its peak, the British Empire, comprised colonies in North America, Australia, Africa, and South Asia. In 1914, around 400 million people and 33,000 km^2 of land lived under British rule. For indigenous societies in Australasia, North America, and southern Africa, British imperialism resulted in violent repression and near-eradication.

Historian Judith Kenny (1995) concisely summarise British imperialism in the Indian Subcontinent in the nineteenth century, following its arrival during the seventeenth century. Initially establishing trading posts reliant upon willing local populations to trade and permit their presence, when the Mughal Empire collapsed and as Britain became embroiled in wars with France during the eighteenth century, the ambition of colonial control was exerted. The East India Company, founded in 1600, would seek to expand its profits from India's wealth and began extracting resources, extracting for sale in Britain rather than engaging in trade with indigenous communities. As Britain became more powerfully present in India, the region was framed at home as a place in need of civilising improvement. The failed Indian Rebellion of 1857 saw Queen Victoria declared sovereign Empress of India, as the 1850s witnessed the emergence of scientific theories of race that categorised non-European peoples as inferior, naturalising their subjugation by colonial forces. As the environs, practices, and beliefs of non-Western peoples were cast as disordered, irrational, diseased, and dangerous, the influence of environmental determinism (see Chapter 1) upon social discourse was palpable in how the landscape would come to be engineered, as explored below.

As Sharp (2022: 64) writes, "colonial knowledge had sought to order the world in a taxonomy of the known", and thus, "the engineers of the colonial landscape sought to order the colonies into a knowable pattern". The "traditional" settlements of colonised peoples became objects of curiosity and fear, especially with concerns of their harbouring violent unrest or disease outbreaks. In his own writing, Franz Fanon (1965: 30) offers a striking, succinct summary of the imagined geography of the "native" urban landscape as appearing before colonial eyes:

> The town belonging to the colonised people [...] is a place of ill fame, peopled by men of evil repute. They are born there, it matters little where or how; they die there, it matters not where nor how. It is a world without spaciousness; men live there on top of each other, and their huts are built one on top of the other [...] The native town is a crouching village, a town on its knees, a town wallowing in the mire.

Such environments contrasted the emergent European ideas of rational, ordered urban space, exemplified in town planning from the latter eighteenth century. They were cast as in need of clearing or reorganisation, given increased fears of violent insurgency following the 1857 Indian Rebellion, and concerns over disease in the wake of a series of cholera pandemics during the century. As Foucault (1975) has elaborated within his later work examining the management of life by the state, wider streets, city walls, and the regulating of the circulation of goods and people in and out of urban areas represent long-standing spatial tactics for controlling the spread of disease, facilitated through the built environment. The British therefore embarked on a series of projects to radically rebuild urban centres in India with a more ordered layout conducive to surveillance, sanitation, and the expression of British power and values via monumental architecture and statuary. Thus, urban landscapes operated as "social technologies" seeking to "incorporate, categorise, control and reform" those living under imperial rule (Sharp, 2022: 64).

Either by destroying non-European settlements, or constructing entirely new "European quarters", such projects of landscape organisation reflected the range of Orientalist assumptions about the landscapes of the other, and the colonisers "civilising role". An example: historical geographer Stephen Legg examines the construction of New Delhi during the early twentieth century. In contrast to the existing "old town" this new landscape, intended as the colonial capital, would comprise open, gridded streets, grand neoclassical buildings, and war memorials celebrating not British oppression but the Indian contribution to imperial military success. The city would frame India as part of a family of nations via its very appearance, as well as better enabling residents to be categorised, sanitised, controlled, and reformed. As Legg describes, seeking to address issues of overcrowding within so-called slums, the British hoped to simultaneously engineer a better society:

> [Whilst] overcrowding was associated with various diseases, from tuberculosis to scabies, smallpox and measles, it also had a bad "social effect". [...] Slums acted on the health and habits of the

> people and encouraged a lassitude of mind that reacted upon the body, which hit people's "resisting power", and thus encouraged immorality, intemperance, gambling and other rampant vices. As such, the need for intervention in slums was not just biopolitical, in terms of health, but governmental, in terms of conduct and moral self-regulation.
>
> (Legg, 2006: 720)

Thus, the landscape in imperial India, as in other colonial contexts, served as a vehicle for ruling power. Whether in towns, plantations (below), or even highland hill stations (see Kenny, 1995; Baker, 2014), the environment was organised to assist in the disciplining of its inhabitants.

As "social spaces", imperial landscapes were certainly characterised by "highly asymmetrical relations of domination and subordination" (Pratt, 2008: 7). Importantly, however, they were not simply sites where indigenous people and environments were remade on the terms of British or other interests. As "the space of imperial encounters [...] in which peoples geographically and historically separated come into contact with each other", such landscapes opened onto various "ongoing relations [...] usually involving conditions of coercion, radial inequality, and intractable conflict" *as well as* relationships that were "interactive" or "improvisational", and signified a different kind of "co-presence" (Pratt, 2008: 8). Furthermore, the question of how colonised peoples resisted, acted back, adopted, or adapting to European efforts at exerting control and meaning upon the landscape sit central within a fulsome understanding of imperialism and landscape within which there is scope for nuance and an appreciation of diverse forms of agency (see also Chapter 7). We consider some of these points via the plantation example.

The Plantation

An infamous "landscape of empire", plantations involved the turning over of an area to intensive cultivation of a single, often commercial, crop via monotonous, mechanical, and the labour of largely enslaved peoples. As anthropologist Sophie Chao describes, the model for such landscapes was established in Europe in the fourteenth century, before spreading overseas as part of "racialized colonial modernity". Cotton and rice were grown in the Americas, whilst coffee, palm, and other crops were grown in places such as India and Indonesia across plantation sites. Thus, the plantation names both a "material formation and conceptual analytic", expressing and encapsulating as it does the violence of colonial projects (Chao, 2022: 361).

For historian Judith Carney: "colonial era plantations so radically shaped and altered landscapes that they can be said to epitomize a global epoch that is still dominated by their land use legacies" (2021: 1075). Indeed, today, plantations endure and proliferate, primarily (though not exclusively) as landscapes within the "Global South" growing food, materials, and fuels for export and consumption within (mostly) wealthier countries. Landscapes marked by "silence and singularity" (Chao, 2022: 361), absent of lively biodiversity, they represent "the principles of ecological simplification, homogenisation, and instrumentalization" that characterise contemporary as much as imperial-age capitalism's relationship with nature. They are also places animated by emergence and resistance (discussed below) and the unruly ways in which more-than-human agencies assert themselves underscores the inherent hubris of such attempts to impose homogenous ecosystemic conditions (see Tsing, 2015).

In their layout, plantations also exemplify – as, with many other imperial landscapes – the operation of *softer* forms of power. Open, organised, and with rows and walkways arranged at regular intervals to enable easy surveillance of workers, the plantation (like the re-engineered colonial capital – see Box 6.2) can be considered with reference to Michel Foucault's (1975) discussion of the panopticon (Figure 6.2). This ideal prison structure, proposed in the eighteenth century by British moral philosopher Jeremy Bentham, would be circular, with prisoners housed such that their rooms faced inwards, towards a central guard tower

Figure 6.2 Interior of Presidio Modelo, a model prison facility on Isla de la Juventud, Cuba. Constructed between 1926 and 1931 and operating until 1967, the prison's panopticon design allowed guards within the central tower to maintain constant surveillance of around 1,000 prisoners in the surrounding cells. Image by Friman. Wikimedia Commons, CC BY-SA 3.0.

commanding a 365° view over the cells all around. The premise underlying the design was that, at any one time, no prisoner within the structure could be sure that they were not being observed. Consequently, prisoners would police *themselves*, always acting as though they were being watched, and therefore internalising the disciplinary norms of the space. For Foucault, the panopticon provided an effective metaphor to understand the operation of surveillance and self-governance as part of a modern, softer form of power in society. In turn, the colonial urban landscape, as described by Legg (2006), or the plantation, as described by James Duncan (2002), has been likewise understood on such terms; being laid out to maximise the visibility of inhabitants to imperial authority and thus instil a self-discipline that would cultivate moral improvement.

Plantation landscapes were imbued with violence. The kind of labour they sustained, and which sustained them, was fundamental to the workings of the colonial economy. For theorist Achille Mbembe (2019), plantations exemplify what he terms the "necropolitics" of Empire. That is, the sustaining and flourishing of society, economy, and everyday life in the European centre required violence and death to be committed elsewhere. Such landscapes underlie (at times, quite literally) wider networks of imperialism and associated spaces, as Katherine McKittrick (2013) demonstrates in her reflections on the New York African American Burial Ground. Excavated beneath Lower Manhattan in 1991, the site interred between 10,000 and 20,000 enslaved people during the seventeenth and eighteenth centuries. For McKittrick, this space testifies to the violence and death experienced by Black enslaved peoples, and the way that legacy comprises "part of the environment we presently inhabit" (2013: 2). Moreover, its legacies are reflected and continued through a range of other spaces, such as the prison and marginalised urban neighbourhood, defining contemporary "Black geographies" (see Chapter 7) and continuing to express the logic of the plantation as both a landscape of violence *and* "innovative resistances" (McKittrick, 2013: 5).

In this latter sense, plantations have proven valuable landscapes for illuminating the *hubris* of Empire; the failure of total domination, and the varied ways in which the agency of colonised, enslaved peoples found expression. For example, Carney examines the role played by Black enslaved peoples in introducing strains of rice, and methods of their cultivation, from Western Africa (see Watkins and Carney, 2022). She has also considered the "slave subsistence plot" and "dooryard gardens" outside slave dwellings, wherein Black enslaved peoples cultivated food (often during time that had to be struggled for) to sustain themselves amidst the brutal conditions of the plantation (Carney, 2021). These practices and histories can reveal plantation space as contested and negotiated; slave gardens contribute to a distinct, mosaic at their edges, supporting "an alternative food system that developed within the interstices of plantation landscapes" (Carney, 2021: 1076). Introduced crops (such as yams and millet), alongside traditional agricultural knowledge on the part of West African enslaved peoples saw these gardens serve as a precursor to the present-day "agrobiodiversity" and small-plot farming that endures in the Caribbean. Thus, plantation landscapes were environments shaped by more than just white Europeans and slave-owners. They were the site of an emergent "subaltern food system" by which various now taken-for-granted ingredients would later arrive upon white dinner tables (2021: 1079).

Likewise, James Duncan's (2002) work examining coffee plantations in imperial Ceylon (Sri Lanka) sees him read "against the grain" of a British-authored archive, comprising documents such as governors' or planters' reports, to recover various tactics of covert and overt resistance employed by Tamil labourers within these landscapes. Such tactics included feigning illness, hiding on site (including amidst vegetation, deploying a local geographical or ecological knowledge to outwit overseers), sabotaging equipment, stealing from stores to meet coffee quotas, and even outright strike action or insubordination. Deployed in the face of threatened or actual pay cuts, beatings, and efforts to organise plantation landscapes to make them easier to survey (Duncan, 2002: 326–329), these practices endure as part of plantation histories. Furthermore, as Duncan considers, the power and race-relations defining these landscapes were complicated by the involvement of some Tamil labourers as overseers or recruiters, both assisting and hindering (for example by beating their charges or feeding them poorly) the productivity of European plantations. Thus, attention to the multifaceted histories of these landscapes reveals a more complex, contested, and interactive version of imperial space than might be assumed.

Empire and Landscape in the Metropole

As Archive Davies (2021: 741) observes, attention towards the infrastructural developments that have defined urban landscapes over recent centuries illuminates the "networks" by which "colonialism and imperialism operated"; as well as the ways in which such infrastructures continue to shape places and their interconnections for the future. Before concluding this chapter, it is important that we acknowledge that Empire was not something that happened *somewhere else.*

In a material sense, the landscapes of nations such as Britain were evidently transformed with the arrival of new resources, non-human species, and people who, often under duress, travelled there from the colonies. Consider the emergence of zoos during the nineteenth century: a material reorganisation of the urban landscape to accommodate charismatic animals gathered from across the Empire, such that they might be put on display to celebrate or educate the public (see Anderson, 1995). In a similar vein, botanical gardens – such as those established in cities like Paris or Edinburgh, as well as the new capitals of the "neo-tropics", such as Christchurch in Aotearoa New Zealand – exhibited the variation and wondrous diversity of plant life from across an Empire, whilst also fulfilling a colonial knowledge project of specimen collection, classification, and categorisation of the world in terms of European/Western ideals (see Ginn, 2008).

Moreover, the wealth extracted from overseas ventures contributed to massive building projects too, transforming the metropole (the ruling state of a colony) into an appropriately imperial centre. Indeed, as a conceptual project, imperialism demanded transformations of so-called home landscapes in order that they reflect the ideals the conquering nation held about itself and its place in the world. Thus, imperialism was involved in transformations at *home* as in distant places, with implications for landscape, its appearance, and associations. Attention to this is central when seeking to understand how imperialism "makes the world meaningful to its subjects" (Pratt, 2008: 3).

The Imperial City of London

The realities of Empire as a process that unfolded in various colonies *and* "happened in the minds and practices of people within Europe as well" (Gilbert and Driver, 2000: 24) becomes especially clear when considering the emergence of Imperial urban landscapes from the mid-latter nineteenth century in cities such as Paris and (as discussed here) London. As Felix Driver and David Gilbert describe with reference to London, the mid-nineteenth century saw an emerging set of discourses and narratives about the "Imperial City"; its ideal form and function as the hub of Empire-building activity. Although London, like many other European capitals, had a significantly longer historical status as imperial capital, there was much anxiety at this time about that city's capacity to compete in terms of physical grandeur with cities on the continent such as Paris. Yet, although such cities competed with each other for status in some senses, ideas and approaches to urban landscape design were frequently exchanged between them as architects communicated and conversed at exhibitions and in correspondence. Consequently, a common form to the imperial capitals emerged through shared influences and the adoption of similar styles.

In particular, the mid-nineteenth century saw the emergence of an "imperial style" of architecture that drew heavily on classical motifs and tropes, including columns and the use of statuary. By recreating London in the image of an imagined, golden-era Rome, the urban landscape would evoke past greatness, whilst also asserting the British Empire as the spiritual successor to that of Rome. Classical motifs signified a timeless superiority and rationalism of the coloniser, in stark contrast to the backward, exotic, and barbarous spaces of the colonised. New designs unveiled for the Foreign and War Offices in Whitehall in the wake of the 1855 Paris International Exhibition saw architects emulating many aspects of recent, classically-styled French buildings. Elsewhere, the Foreign and Colonial offices, completed in 1874, were adorned with the busts of famous British explorers, such as Captain Cook and David Livingstone.

As Driver and Gilbert also discuss, imperial landscapes were sites for *performing* Empire rather than static accumulations of symbols. Via exhibitions and public procession, these were "spaces in movement" (Driver and Gilbert, 1998: 14) and practice (see Chapter 8). From the middle of the century, there was much rebuilding around Whitehall (an area of central London where many governmental offices are located). This included schemes to widen and embellish the Mall (Figure 6.3) – a ceremonial route linking Buckingham Palace with Whitehall, terminating at famous Trafalgar Square – including by way of new symbolic structures like "the Admiralty Arch", in accordance with the vision of architect Sir Aston Webb (1849–1930). These proposals aimed not only at enhanced grandeur but also created a

Figure 6.3 The Mall looking north-west towards the Admiralty Arch from Buckingham Palace, London, UK. Image by Gusim. Wikimedia Commons, CC BY-SA 3.0.

landscape to host extravagant performances of imperial power. Webb's designs created a "triumphant axis" linking Whitehall with the newly refurbished Palace in the early twentieth century: an urban landscape of monumental structures, parades, and pageantry worthy of "the heart of empire" (Driver and Gilbert, 1998: 12).

Other sites in London hosted performances of different kinds. In South Kensington, museum displays and live exhibitions throughout the nineteenth century showcased the vast extent of territorial control and the resources and treasures this yielded for Britain. Moreover, tableaux and displays packaged up stereotypical representations of non-Western cultures in ways that emphasised the fundamental difference of the Orient. As Sharp (2022) describes, at some such exhibitions, small villages were even reconstructed and staffed with indigenous peoples, taken to the metropole, to enable those at home to literally walk amidst a microcosmic version of the native landscape and observe its strangeness.

Pause for thought: *Take a moment to consider the phenomenon of the imperial exhibition. How might such an event be received today? Where might we encounter the legacies or remainders of this imperial exhibitionist culture in contemporary landscapes?*

Aside from spaces specifically *designed* to express imperial discourse, the city landscape materialised the Empire in other ways. In London's Strand, the embassies and offices of the imperial dominions – including Australia House and India

House, built in 1918 and 1930 respectively – gave distant colonies (and their government infrastructure) a presence in the metropole. Likewise, visiting colonial dignitaries, proto-tourists, and migrants, passing through the city, constituted another way in which the Empire manifested at home. These material traces, then, evidence the influence of the Empire upon the physical and conceptual formation of metropole cityscapes.

Chapter Summary

Box 6.3 Key Concepts

Chapter 6

Empire
Imperialism/Colonialism
Imagined Geography
Contact Zone
Metropole

This chapter has examined the historical relationship between imperialism – the project of empire-building and its associated power relations, material implications, and cultural dimensions – and the production of landscapes (Box 6.3). The examples discussed constitute a mere fraction of the sites (and histories) marked by Empire. Nevertheless, they do illuminate key lines of connection between landscapes and imperialism. Indeed, focusing upon landscape offers a means for imperial histories to attend to the physical impacts of colonial activities alongside an analysis of the discourses that propagated (and were reproduced) through such activity. Understanding the consequences of Orientalism, or the mobilisation of myths about a pristine "New World", becomes possible when one considers discourse and its effects via specific landscape case-studies.

Moreover, and though beyond this chapter's scope, no less important are those new forms of imperialism, or "neocolonialism", that constitute the various forms of a "colonial present" (Gregory, 2004). For instance, scholars have noted how Western funding, discourses, and institutions of conservation have colonised numerous non-Western landscapes, with implications for how those places are valued and perceived, as well as how they are materially organised (potentially against the interests or values of indigenous communities). Thus, familiar power imbalances and lines of difference are (re)drawn, be it through ways of representing wildernesses in Mexico (Rasmussen, 2024) or "landscaping" a new colonial present in Tanzania (Bluwstein, 2021), in ways that emphasise how attention to Sluyter's "manifold" of landscape and imperialism remains relevant today. Indeed, as the following chapter explores, whilst imperial influences on "landscape" linger on, a great many alternative landscape ontologies also abound, challenging any notion that European perspectives on the world are the only perspectives of note.

A challenge implicit to much of the scholarship that informs this chapter, as explicitly discussed in Chapter 7, concerns the often-limited available information on how imperial landscapes were experienced, understood, or resisted by colonised peoples. Moreover, limited (written) evidence documents the kinds of landscape practices and meanings associated with pre-colonisation indigenous societies. Much research exploring landscapes of Empire, at home or abroad, relies upon archival records produced during (or as part of) imperial rule: "the colonial archive" or "imperial archive" (see Duncan, 1999; Craggs, 2008). Such materials tend to represent a worldview that takes Empire, and its ways of seeing, thinking about, and knowing the world, for granted. Other ways of knowing the world that do not fit with this framework, or which fall outside of European norms, are more rarely documented. Moreover, many indigenous communities encountered by Europeans had cultures reliant more on oral and artefactual modes of knowledge transmission and tradition. Thus, as Sluyter observes, whilst there is a need for "relevant data to operationalise the colonial triangle" (2001: 424) and understand relationships between coloniser, colonised, and landscape, the recovery of non-European agency in these relations can prove difficult in the absence of adequate historical sources.

Nevertheless, creative engagement with existing archival resources, as well as the use of alternative sources (including, but hardly limited to, comparative study of landscape morphology, engagement with archaeological remains, pollen cores or other organic and inorganic deposits, folklore, oral histories, and recorded cosmologies or belief systems) *can* offer means to posit alternative stories and practices of landscape (see Watkins and Carney, 2022). Some of the examples discussed above reflect such efforts; whether examining fire regimes in Spanish-colonised Mexico by way of environmental data, or identifying acts of Sri Lankan plantation workers' resistance from reading available sources differently and attentively. As Pratt notes, "[o]ften the European archive provides valuable raw material on which to exercise a decolonizing creativity" (2008: 8). There is, as Gayatri Spivak (1988: 211) articulates, an incentive to "read against the grain" of existing sources and follow "misfits in the text". Seeing absences or elisions as a starting point for recovering, or illuminating, something of the colonised subject's experience and contribution to the making of landscape, draws us into the crafting of different kinds of landscape studies, and new ways of working with sources.

Suggested Seminar Activities

Activity 1: Explore a landscape in your local area and consider how this landscape might be understood to have been produced through the relationships of "imperialism".

- How has it been materially produced through the commercial, labour, cultural, economic, or ecological relationships of Empire?
- How has the way this landscape is perceived, or designed to be perceived, and invested with meaning for you or others, been affected by the processes of imperialism?

Activity 2: Gather a selection of tourist brochures or websites.

- How are distant places still being viewed through the language or imagery of Orientalism?

Recommended Reading

Carney, J. 2021. Subsistence in the Plantationocene: Dooryard gardens, agrobiodiversity, and the subaltern economies of slavery. *The Journal of Peasant Studies* 48(5): 1075–1099.

Davies, A. 2021. The coloniality of infrastructure: Engineering, landscape and modernity in Recife. *Environment and Planning D: Society and Space* 39(4): 740–757.

Sluyter, A. and Duvall, C. 2016. African fire cultures, cattle ranching, and colonial landscape transformations in the neotropics. *Geographical Review* 106(2): 294–311.

References

Anderson, K. 1995. Culture and nature at the Adelaide zoo: At the frontiers of 'Human' geography. *Transactions of the Institute of British Geographers* 20(3): 275–294.

Arnold, D. 1993. *Colonizing the Body: State Medicine and Epidemic Disease in Nineteenth-Century India*. Berkley: University of California Press.

Arnold, D. 2000. "Illusory Riches": Representations of the tropical world, 1840–1950. *Singapore Journal of Tropical Geography* 21(1): 6–18.

Arnold, D. 2006. *The Tropics and the Traveling Gaze: India, Landscape, and Science, 1800–1856*. London: University of Washington Press.

Baker, J.C. 2014. Mobility, tropicality and landscape: The Darjeeling Himalayan Railway, 1881–1939. *Journal of Historical Geography* 44: 133–144.

Bluwstein, J. 2021. Colonizing landscapes/landscaping colonies: From a global history of landscapism to the contemporary landscape approach in nature conservation. *Journal of Political Ecology* 28(1): 904–927.

Carney, J. 2021. Subsistence in the Plantationocene: Dooryard gardens, agrobiodiversity, and the subaltern economies of slavery. *The Journal of Peasant Studies* 48(5): 1075–1099.

Chao, S. 2022. Plantation. *Environmental Humanities* 14(2): 361–366.

Colás, A. 2007. *Empire*. Cambridge: Polity.

Cosgrove, D. 2003. Landscape and the European sense of sight: Eyeing nature. In: Anderson, K., Domosh, M., Pile, S., and Thrift, N. Eds. *Handbook of Cultural Geography*. London: Sage, pp. 249–268.

Craggs, R. 2008. Situating the imperial archive: The Royal Empire Society Library, 1868–1945. *Journal of Historical Geography* 34(1): 48–67.

Crosby, A. 2015. *Ecological Imperialism*. 2nd Ed. Cambridge: University of Cambridge Press.

Davies, A. 2021. The coloniality of infrastructure: Engineering, landscape and modernity in Recife. *Environment and Planning D: Society and Space* 39(4): 740–757.

Driver, F. and Gilbert, D. 1998. Heart of empire? Landscape, space and performance in imperial London. *Environment and Planning D: Society and Space* 16(1): 11–28.

Duncan, J. 1990. *The City as Text: The Politics of Landscape Interpretation in the Kandyan Kingdom*. Cambridge: Cambridge University Press.

Duncan, J. 1999. Complicity and resistance in the colonial archive: Some issues of method and theory in historical geography. *Historical Geography* 27: 119–128.

Duncan, J. 2002. Embodying colonialism? Domination and resistance in nineteenth-century Ceylonese coffee plantations. *Journal of Historical Geography* 28(3): 317–338.

Fanon, F. 1965. *The Wretched of the Earth*. London: Penguin.
Foucault, M. 1975. *Discipline and Punish: The Birth of the Prison*. London: Penguin.
Gilbert, D. and Driver, F. 2000. Capital and empire: Geographies of imperial London. *GeoJournal* 51: 23–32.
Ginn, F. 2008. Extension, subversion, containment: Eco-nationalism and (post) colonial nature in Aotearoa New Zealand. *Transactions of the Institute of British Geographers* 33(3): 335–353.
Greer, K.A. 2020. *Red Coats and Wild Birds: How Military Ornithologists and Migrant Birds Shaped Empire*. Chapel Hill, NC: UNC Press Books.
Gregory, D. 2004. *The Colonial Present*. Oxford: Wiley Blackwell.
Hunter, R. and Sluyter, A. 2011. How incipient colonies create territory: The textual surveys of New Spain, 1520s–1620s. *Journal of Historical Geography* 37(3): 288–299.
Kenny, J.T. 1995. Climate, race, and imperial authority: The symbolic landscape of the British hill station in India. *Annals of the Association of American Geographers* 85(4): 694–714.
Legg, S. 2006. Governmentality, congestion and calculation in colonial Delhi. *Social & Cultural Geography* 7(5): 709–729.
Mbembe, A. 2019. *Necropolitics*. Durham, NC: Duke University Press.
McKittrick, K. November 2013. Plantation futures. *Small Axe* 42: 1–15.
Mitchell, W.J.T. 2002. Imperial landscape. In: Mitchel, W.J.T. Ed. *Landscape and Power*. 2nd Ed. London: University of Chicago Press, pp. 5–34.
Phillips, R. 2014. Colonialism and post-colonialism. In: Cloke, P., Crang, P., and Goodwin, M. Eds. *Introducing Human Geographies*. 3rd Ed. Abingdon: Routledge, pp. 493–508.
Pratt, M.-L. 2008. *Imperial Eyes: Travel Writing and Transculturation*. 2nd Ed. Abingdon: Routledge.
Rasmussen, B. 2024. (Un)making a conservation landscape: Repeat photography and environmental narrative in Mexico's Sierra de San Pedro Mártir National Park. *Landscape Research* (*online* 14 November 2024). https://doi.org/10.1080/01426397.2024.2428327
Safier, N. 2014. The tenacious travels of the torrid zone and the global dimensions of geographical knowledge in the eighteenth century. *Journal of Early Modern History* 18(1–2): 141–172.
Said, E. 1978. *Orientalism*. London: Penguin.
Sharp, J. 2022. *Geographies of Postcolonialism: Spaces of Power and Representation*. 2nd Ed. London: Sage.
Sluyter, A. 1999. The making of the myth in postcolonial development: Material-conceptual landscape transformation in sixteenth-century veracruz. *Annals of the Association of American Geographers* 89(3): 377–401.
Sluyter, A. 2001. Colonialism and landscape in the Americas: Material/conceptual transformations and continuing consequences. *Annals of the Association of American Geographers* 91(2): 410–429.
Sluyter, A. and Duvall, C. 2016. African fire cultures, cattle ranching, and colonial landscape transformations in the neotropics. *Geographical Review* 106(2): 294–311.
Spivak, G. 1988. Subaltern studies: Deconstructing historiography. In: *In Other Worlds: Essays in Cultural Politics*. London: Routledge, pp. 197–221.
Tsing, A. 2015. *The Mushroom at the End of the World: On the Possibility of Life in Capitalist Ruins*. Princeton, NJ: Princeton University Press.
Watkins, C. and Carney, J. 2022. Amplifying the archive: Methodological plurality and geographies of the Black Atlantic. *Antipode* 54(4): 1297–1319.

7 Decolonising Landscape

Introduction: On Decolonising Landscape

In his collection, *The Cartographer Tries to Map a Way to Zion* (2014), the Jamaican poet, writer, and literary scholar Kei Miller dramatises an encounter between a Western cartographer, seeking to map the island as “territory”, and the “rastaman”, a sometime cipher for Miller himself, who responds, challenges, and confounds this project by way of lyrical exchanges, sermons, and meditations on place, history, and memory. The cartographer, attempting to account for the island landscape, finds himself confronted and confounded by its environment, and the ways in which its inhabitants live and know it. His efforts to follow roads, ridges, hills, and shorelines see him attempt to dispassionately document the land, refusing to fall in “love” or get involved with its “muddy affairs” (2014: 18). Responding, the rastaman chides these efforts, “to make thin and crushable all that is big and real” (2014: 17) and taunts the possibility of countering any *official* mapping with alternative tracings of his own.

This tension between coloniser and colonised, and their ways of seeing landscape, is encapsulated in Miller’s poem, “What the Mapmaker Ought to Know”. Miller describes an island environment where “things fidget”. The landscape he conjures “does not sit / willingly / as if behind an easel” awaiting efforts at its categorisation. Indeed, it is more likely “to slip / out from under your grasp” (Miller, 2014: 15). In this way, the poems, lyrics, and speculative dialogues in Miller’s collection offer encounter with different versions of the Jamaican landscape: both “a palimpsest of maps” (2014: 25), more-than-human and connected to other places and times; and a site that resists (or *un*-maps) the very act of mapping. The land’s history, cosmology, and (oft suppressed) rhythms surge forth to reassert those things excluded from the map (see poem “In Praise of Maps” – Miller, 2014: 63). In this way, landmarks, place names, and *landscapes* reject attempts at ordering and serve as openings, for Miller, into old and new stories, lived relations, island specificity, and the enduring violence of colonisation: what was taken, what was given, what is owed, and what remains. The implications for landscape’s study and meaning are encapsulated in the poem above: colonial modes of thought cannot do it, or its inhabitants, justice.

DOI: 10.4324/9781003172451-8

In Chapter 6, we outlined the relationship between Empire and landscape as co-constituting, historical, and ongoing as environments, which were reconstructed in the name of imperial progress, order, and civilisation in the metropole and overseas (Mignolo and Walsh, 2018). Here, we attend to the fact that, more than a material or imagined environment, landscape bears the marks of the imperial project *as both concept and project of study*. Scholars of landscape (as many other disciplines besides) must consider their implications in *continuing* the imagined and material structures, inequalities, and violence of Empire. As geographer Sarah Radcliffe remarks, our world is characterised by "relations of power, organisations of space, priorities, and mindsets that [...] originated in modern forms of colonialism and empire" (Radcliffe, 2022: 1). Landscape studies are no different. Such scholarship often echoes the "unbearable whiteness" exuded by much of the social sciences, humanities, and natural sciences (Derickson 2017: 236). The subject areas wherein landscape studies occur, such as geography, arguably continue to demonstrate "little practical contemporary openness to difference and diversity in its knowledge production processes" (Noxolo, 2017: 317).

Therefore, as empirical landscapes can be studied in relation to Empire, so too can landscape scholarship's relation to imperialism be subject to interrogation. Doing so is vital if we are to ask what it means to decolonise the study of landscape, as this chapter attempts to do, in five key ways. Firstly, we address the implications of landscape scholarship in the historical processes and production of the Empire. Secondly, we consider how landscape carries within it a set of (Western) assumptions about the world (and our relationship to it) that require "unlearning". Thirdly, we explore some methodological and epistemological strategies proposed to assist this unlearning. Fourthly, we highlight how decolonial landscape study requires engagement with landscapes both near and far, including those within the so-called "heart" of Empire. Finally, we look at how advancing decolonial *scholarship* of landscape might translate into material acts of decolonisation, including the reassertion of control and authority over land.

This chapter brings together a range of work conducted under the auspices of postcolonialism, decolonial studies, and Indigenous ontologies. We also acknowledge – though do not necessarily engage substantively with – the significance of Black studies and its take-up within areas including Black geographies (see Box 7.1) as part of this tradition. Labels such as those above are in active use and flux. Defining them is "notoriously difficult", yet necessary, given each indicates a body of work opposed to essentialising experience into the logic of the coloniser (see Jazeel, 2019). For our purposes here, we follow Tariq Jazeel's useful overview to understand both postcolonial and *de*colonial scholarship as that which considers how the legacies of historic colonialism and imperialism continue to shape and reproduce the world in which we live. This includes an attention to how particular ways of thinking, perceiving, and acting upon landscapes, peoples, and cultures are evaluated (and valued) in terms of *difference*. This work is thus concerned with enacting "an intention towards pluralisation, towards spatial difference" (Jazeel, 2019: 2) by means of "its insistence on taking particularity and context seriously" (2019: 14). However, decolonial work, in contrast to the postcolonial, asserts a need to actively engage in dismantling the residual structures of Empire, and challenges any notion that we exist *after* imperialism.

Box 7.1 Black Geographies

The field of "Black geographies" might be located within a broader corpus of decolonial and race scholarship, albeit expressing a specific concern with "the geographical expressions of Black populations" (Bledsoe and Jamaal Wright, 2019: 419), especially with a view to realising Black liberation. As Hawthorne writes:

> Black Geographies is not exclusively the study (or rather, positivist description) of Black people, nor does it entail the identification of some sort of reductive, non-relational "Black space." It is also not simply a matter of adding Black subjects to geography syllabi in a liberal multicultural project of "add diversity and stir." Rather, the use of "Black" in the phrase "Black Geographies" is intended, at least in part, to serve as a radical provocation to the discipline of geography. It is a call to center those subjects, voices, and experiences that have been systematically excluded from the mainstream spaces of geographical inquiry. It is also an invitation to consider how an analysis of space, place, and power can be fundamentally transformed by foregrounding questions of Blackness and racism.
>
> (2019: 8)

Such work reckons explicitly with how different spaces, places, and landscapes are, in myriad ways, coded as *anti*-Black, or organised around a norm of non-Blackness. Black scholars have therefore worked to engage and articulate "Black spatialities", and contest the enduring "anti-Black histories" and the "traditional" logics of disciplines such as Geography (see Noxolo, 2022; also McKittrick, 2006), with a view to countering their (re) production and offering alternatives. More radically, Black studies and Caribbean thought raise fundamental questions about the nature of the subject and their world (see Pugh and Chandler, 2023). Such scholarship has grown both from a longer historical effort by Black scholars and radical thinkers to articulate Black, diasporic and otherwise marginalised or excluded groups' spatial experience, practices and politics (see Hawthorne, 2019) and more recent explicit (and named) projects initially driven by (African-)American Black scholars such as LaToya Eaves, Katherine McKittrick and Ruth Wilson Gilmore. This latter body of scholarship is defined by an effort to "[critique] and [catalogue] the many ways in which Black lives have been shaped by the spatial logics and conditions of colonialism, enslavement and white supremacy under the unequal environmental and territorial conditions of racial capitalism" (Noxolo, 2022: 1233); and to counter the persistent "omission and erasure" of Black thought within a "white-dominated global knowledge" (2022: 1234).

Whilst such work has not explicitly engaged with landscape – partly owing to the desire to operate outside of such "traditional" concepts, steeped in colonial thought – there are clearly opportunities for landscape studies to engage Black Geographies. As reviews by Hawthorne (2019) and Noxolo (2022) outline, key areas within Black Geographies scholarship include: the legacy of slavery and its spatial formations; Black communities and individuals' sense of place; the centrality of race and racism to the workings of capitalism; and the experiences of Black urbanism and the violent structures of the criminal justice system. As the previous chapter demonstrated, attention to plantation landscapes (after McKittrick, 2011) enriches consideration of the relationship between Empire, landscape, violence, and modernity. Certain key Black thinkers also engage with landscape. Edouard Glissant – influential writer and theorist of creolization – articulates the landscape of Martinique as something that "speaks" to and informs our being-becoming through encounters and moments of "fusion" with difference (or its opacity) (see Britton, 2014). Finally, key Black concepts like marronage – and the alternative spatial arrangements these suggest – have been applied in recent work to explore the lasting impact of colonial environmental histories on contemporary arboreal landscapes (Moulton, 2023).

We also distinguish, as others, between metaphorical decolonial approaches and decolonising praxis. The former aims at dismantling colonial ways of thinking and knowing as they sustain relations of domination. The latter asserts more actively questions of political re-enfranchisement and the restoration of rights in an avowedly practical and nonmetaphorical sense (see Tuck and Yang, 2012; Dang, 2021). Metaphorical approaches chime with debates about the need to decolonise education or represent more diverse accounts, histories, and outlooks on the world. A praxis approach concerns both efforts to highlight and seek compensation for institutional involvement in colonial atrocities (e.g. slavery reparations); as well as engagement with contemporary Indigenous struggles over landownership, rights, and territory. Whilst this chapter, being concerned with the *study* of landscapes, mostly focuses on the former, we argue that such activity cannot be separated from the latter, as we seek to address in the final section. Furthermore, such approaches need not exist in isolation, and may well percolate into one another.

A final note: this chapter features in a textbook written by two white-British academics whose own research interests concern landscape, nature, and (primarily) the Scottish Highlands. We have tried to offer as useful an account as possible to introduce landscape scholars to the vital and important work of decolonising landscape research, but it remains, inevitably, a partial work of two outsiders to these debates. We hope and encourage readers to seek the works cited (and the works they in turn reference) and engage directly with this material themselves. Neither an attempt to define nor set agendas for this field, the chapter is merely a springboard into discussions for those wishing to get started.

Chapter Outline

This chapter examines some key issues regarding how to decolonise landscape research. We begin by highlighting the imbrication of "landscape" as a concept with histories of imperialism, before questioning how such ways of thinking can be "unlearned". We also consider the credibility (or appropriateness) of "landscape" as a term, and conventional ways of framing its analysis, when applied in non-Western contexts. We reflect on how scholars have often engaged with Indigenous knowledge systems to craft accounts of human-environmental relations that reject some of the assumptions freighted by "landscape"; the ways in which decolonial landscape work might explore the environments both in former "colonies" *and* the metropole. Finally, after Tuck and Yang (2012: 36), we posit decolonisation as an "elsewhere" that must be striven towards, outlining two examples of how, practically and materially, decolonisation prompts different ways of organising and managing *land.*

Unlearning Landscape

The previous chapter considered how the landscape was perceived, organised, and known under the Empire. It is, however, also the case that the *study* of landscape (within scholarship), its definition (in theory and academic debate), and those disciplines providing the means for its examination (history, literature, sociology, anthropology, geography, and others) are likewise embroiled with legacies of Empire. Indeed, much landscape scholarship draws upon a legacy that is predominantly White and Euro-American. Consequently, the assumptions of "Western" cultures, their ways of conceiving human-environmental relations, model the persistence of colonial discourses. Uncritically deploying Euro-American ways of thinking performs (if even unwittingly) a colonial act of erasure by which alternative knowledge systems and perspectives are ignored or marginalised. The imperial knowledge project is thus perpetuated as forms of domination, oppression, and (epistemological) violence remain unchallenged.

Consider geography, the authors' primary discipline. The discipline was a handmaiden of Empire, and much has been written as to the production of geographical knowledge (maps and accounts of different landscapes, environments and their inhabitants), emergence of institutions (such as the Royal Geographical Society), and codification of practices (including mapping) in support of the Imperial project (see Driver, 2001). For example, J.B. Harley's (2001: 46) classic exploration of the map as "predominantly a record of colonial self-interest" reveals it as a weapon of colonial control and Empire, making distant places available for strategic planning, governance, and subjugation. Thus, like those landscape paintings or literary works that romanticised "tropicality", geographical representations and scholarly accounts likewise reinforced a sense of difference and inferiority. Similarly, geographers (and others) advanced theories that purported to scientifically rationalise differences in relation to the influence of pathological climates, serving in art historian Charmaine Nelson's words, to produce,

> the colonizing worlds as centre [and thus] also justified, rationalized, and defined the world into which Europe violently intruded as colonizable; a process that "othered" geographies already occupied and lived in by a myriad of indigenous peoples and created the imperial center-periphery model.
>
> (2016: 8)

Whilst landscape as a material and imaginative reality was produced under Empire, it was also articulated and developed *as a concept* in this same colonial context (see Chapter 1). The term therefore carries with(in) it a series of Euro-American assumptions about the meanings of culture, nature, and their interaction. Landscape, like the disciplines which studied it, is likewise a handmaiden of Empire that works to reproduce colonial ideas and thus manifest the "colonial present" (see Gregory, 2004).

Consequently, it becomes necessary to challenge what we think of as landscape and the conventional ways of doing landscape scholarship. To begin to consider how we might decolonise landscape and its study, we have to first appreciate how the unthinking application of Western-centric theories of landscape (as outlined across this book) to non-Western contexts serves to reproduce the dynamics of a colonial relationship, erasing difference, specificity, and alternative forms of knowledge. As Tariq Jazeel argues, echoing sociologist Raewyn Connell (2007), applying theoretical approaches advanced primarily within Anglophone scholarship to non-Western landscapes amounts to "methodological projection" whereby these places are viewed only through the existing values, priorities, and debates of an external intellectual context (Jazeel, 2013: 64). The failure to *situate* landscape's own theoretical pillars and conceptual metaphors risks unthinking "epistemological violence" (Jazeel, 2013: 65). Careful consideration of the way in which research is done matters to avoid an account that "ventriloquises" non-Western landscapes, "occluding a range of other things" (Jazeel, 2013: 65).

Pause for thought: *artist Sonia E. Barrett's Dreading the map installation asks the viewer to consider how the Map Room of the Royal Geography Society building in London might be reimagined. This work is one of a series of provocations produced by the Creative Approaches to Race and Insecurity in the Caribbean and UK project. Watch the following film which documents the project and consider the questions posed on the RGS Dreading the Map* *resources, available via the RGS page: www.rgs.org/research/higher-education-resources/dreading-the-map*

- *Watch the film here:* www.youtube.com/watch?v=Wnfd4HRMn3M

To translate meaning requires engagement with the idioms and ontologies characteristic of specific cultural settings. In an Anglo-American context, words like landscape, environment, land, and ecology also carry with them associated scientific perspectives (Bradley and Kearney, 2018). Decolonial work, in turn, must "acknowledge its partiality, its imperfection, its out-of-placeness" when it ventures to engage landscapes "positioned beyond the staging grounds of [...] the concept itself" (Friess and Jazeel, 2017: 18). For Jazeel, this means that decolonial landscape studies ought to practice "a greater degree of fidelity to field contexts" (Jazeel, 2013: 70). Doing landscape study in this way is "hard work" and requires the researcher to think seriously about how their scholarly approach and assumptions may require reworking in the face of "radical modes of alterity" if they are not simply to be subsumed under a Eurocentric sense of landscape. An example of this kind of working is outlined in Box 7.2.

Box 7.2 Doing Postcolonial Landscape in Lunuganga, Sri Lanka

Working within conceptualisations of landscape as textual and affective, geographer Tariq Jazeel (2013) offers an interpretation of the Lunuganga estate, Sri Lanka. As the home and garden of "tropical-modern" architect Geoffrey Bawa, Jazeel pays particular attention to the view from the house across to Cinnamon Hill (a vista available via an online search for the Lunuganga garden tour). He describes the scene:

> Standing at the house, looking across the thick lawn and up the gentle hill slope, one's view is framed by trees on either side and in the middle distance, on the crest of Cinnamon Hill, a lone moonamal tree looms over a large pot. The tree points to the gleaming white dome of the Katukuliya temple, a Buddhist dagoba nestled in the verdant vegetation of a hill separated from the estate by a thin sliver of lake.

Connecting this landscape to global discourses of international modernism and design, Jazeel highlights how Lunuganga can be read as a "designed" landscape, encoded and choreographed to reproduce a relationship of ownership (see Chapter 2). Yet, he also makes clear that to end the analysis here would be to fail to adequately account for how this landscape is encountered and reproduced within the social and political context of Sri Lanka.

Bawa, the architect, was "the doyen of architectural modernism in post independent Sri Lanka" (Jazeel, 2013: 67). Influenced by European modernist aesthetics, his design also incorporates distinct elements of Sri Lankan vernacular "well suited to discursive co-option by the anticolonial Sinhala-Buddhist nationalism of post independent Sri Lanka" (Jazeel, 2013: 67). Appreciating this landscape in context alters its meaning: Jazeel highlights Bawa's involvement in postcolonial national building projects for the newly established state. Thus, Lunuganga, its appearance and interpretation, demonstrate the use of architecture to perform a post-independence vision of the nation and its power.

Understanding this landscape in ways exceeding the application of Western frameworks offers "a reappraisal [...] sensitive to such ethnically striated vernacularism" (Jazeel, 2013: 68) and necessitates attention to "particular kinds of politics [as they] emerge from differently textualised human relationships with the environment" (Jazeel, 2013: 65). As a postcolonial, regionally attentive version of landscape study, this example highlights the work that landscapes do to "instantiate particular forms of ethnicized hegemony in the representational practices of everyday life" (2013: 76): in this case, a Sinhala–Buddhist nationalist imaginary.

The Inadequacy of "Landscape"?

In a famous speech, later published, the feminist-activist Audre Lorde elaborated on the importance of considering the histories, enduring exclusionary logics, and responsibility to challenge and offer alternatives to the scholarly frameworks we deploy. Speaking at a conference held at New York University's Institute for the Humanities in 1979, focused on a discussion of American women's experiences, Lorde expresses her frustration at the siloing of Black experience (a single panel was convened on the topic) and the fact that many of the papers given at the event, even those acknowledging non-white or non-heteronormative experiences of womanhood, displayed little effort to actually *think with or through* difference and conceptualise American women's lives in different terms or outside of established social norms. Thus, Lorde lamented a tendency for scholars to wield conceptual and analytical "tools" that had their origins in unjust and unequal ways of thinking without reflection or questioning of their appropriateness. In failing to question their own lack of critique, they remained complicit in constraining the "parameters of change" (Lorde, 2019: 103). In other words, the feminist scholars she addressed seemed to depart from only to return to "traditional" ways of thinking the world (cf. McKittrick, 2006), leaving the very structures of thought that perpetuated the violence and inequalities that affected the experiences of individuals, communities, and cultures "who stand outside the circle" unchanged.

For Lorde, these observations affirmed a need to go beyond "mere tolerance of difference"; or the acknowledgement of difference in a cursory yet ultimately passing fashion. Instead, she wished to abide in the prompt of difference to think creatively: to "spark like a dialectic" (Lorde, 2019: 104). This would be to affirm difference and diversity, rather than disavow or "shed" such qualities in a search for universal accounts of female experience, so as to accumulate new knowledge and construct new forms of community. To really listen to the voices, experiences, and ideas of those "outside the circle" of the normative meant engaging with them in terms of what they offered for an *alternative* way of knowing, rather than as a supplement to existing ways of knowing. As Lorde (2019: 105 – orig. emph.) put it: "*the master's tools will never dismantle the master's house*. They may allow us temporarily to beat him at his own game, but they will never enable us to bring about genuine change".

To return to landscape and decoloniality, if landscape is only ever thought through the lens of traditional, "Western" philosophical, white, largely male-defined ontological or epistemological approaches, then certain individuals, experiences, and worlds are either, at worst, removed from the conversation entirely, or, at best, mediated prior to their inclusion by being treated only in a way that allows them to be slotted into conventional forms of analysis. The vitality and "possibility" that their difference carries, and their capacity to open onto other forms or ways of knowing, are listed. For Lorde, "personal visions" of difference open onto political action – ways of doing things differently. Thus, a tendency to focus upon the same references, reading lists, and approaches shuts down such outcomes from the off: an "evasion of responsibility" or a "cop-out" that continues to exclude alternative perspectives or otherwise ensure they remain bracketed out. It behoves the contemporary scholar of landscape, then, to venture beyond, educate themselves out of traditional ways of thinking, and practice a curiosity that sees encounters with difference as sources of illumination, rather than discomfort.

As a result of reckoning with histories of Empire, exclusion, and landscape scholarship itself, we can appreciate that the "tools" (the terms, ideas, approaches) of landscape enquiry are part of the problem. In accounting for the inadequacies of the term landscape, Bradley and Kearny (2018) highlight that this pervasive concept is potentially antithetical to a decolonial approach, as both it and a series of derived terms (e.g. seascape) are ultimately too partial in their reach. As they elaborate:

> does this conceptual tool really work to tell a wide range of possible stories about human life? Is it fully capable of articulating the relational bonds that define the vast range of people's cultural engagements and enmeshments with their worlds? There is the underlying premise that different cultural perspectives are bridgeable by concepts in the English language, with words like environment, landscape, ecology, sustainable development or co-management, that have their origins within a scientific discourse.
>
> (Bradley and Kearney, 2018: 290)

This discomfort stems from the ontological basis of the landscape concept – its basic assumptions about what exists. In reproducing a modern separation of "nature" and "culture" (see Chapter 1), it has the capacity to at best mistranslate, at worst overlay, ways of understanding place to fit with academic assumptions. The result, when working with (or, more commonly, on) Indigenous communities, is that "Indigenous experiences of the relational world are made sense of through an alien discourse". Thus: "In instances where the word landscape is deployed to describe lands that are not our own, and which have their own culturally prescribed meanings and descriptions, *we must increasingly face the possibility that the term is inadequate*" (Bradley and Kearney, 2018: 290 – emph. added).

Reflecting on the modes of thinking inscribed *by* landscape, Bradley and Kearney observe that "*you whitefellas pull it all apart*". By this, they mean that Western conceptual framings flatten alternative understandings of place by imposing pre-prescribed sets of categories or interpretations. Instead, they, as the example below, suggest alternative concepts are required to understand human relations with surroundings, including notions like the Indigenous-Aborigine "country", which encompasses those things that landscape may exclude (the sea, other non-human beings, oral traditions, dreams). Such a call requires more than simply paying lip service to alternative ontologies. As Radcliffe (2022) argues, there likewise needs to be critical reflection on how Indigeneity is understood and engaged with within research. Landscape scholars should neither cherry-pick nor casually "apply" Indigenous perspectives in ways risking exoticism or essentialising, but take care to situate and engage with such communities and their understandings in the context of their histories, present land relationship, and relationship to how that land is managed or governed.

Doing Landscape Study Differently

If there is some appreciation that the conceptual and practical tools of traditional Western landscape enquiry are part of the problem, how have scholars attempted to do things differently? What might decolonial approaches to landscape look like, and how (if at all) can existing landscape approaches be adapted or *re*-tooled?

Whilst many academic disciplines associated with landscape study have long attempted to understand the world outside of a conventional (white/male) perspectives via postcolonial (Jacobs, 1996; Said, 1978), feminist (Mohanty, 1984; McDowell, 1992), or non-Western theories, the need remains to incorporate perspectives from those ordinarily positioned outside of the spaces and imaginaries of Western knowledge. This is vital if we are to cultivate, after Spivak (2004), the kinds of place-focused, attentive, and critical accounts decolonial work demands. Decolonial approaches have therefore increasingly made room for Indigenous ontologies to cultivate, after Linda Tuhiwai Smith (2021), decolonised methodologies (ways of doing research) and epistemologies (ways of knowing and evaluating knowledge).

For Barker and Pickerill (2020), Indigenous ways of knowing offer much potential for new kinds of "doings". That is, the mobilising of "embodied theories and praxes of liberation" (Daigle and Ramirez, 2019: 79) with research: work that involves researchers as whole beings in active and engaged ways. There are increasing numbers of scholars engaged in practical, collaborative work centring Indigenous communities wherein engagements with those groups "requires an active 'doing' of different ways of being in our professional and personal lives" to both benefit the partners involved and subvert the colonial structures of knowledge production that might serve to exclude such partners, and their interests or outlooks, from the academy (Barker and Pickerill, 2020: 641). As Barker and Pickerill describe, reflecting on their work as geographers, "decolonising [...] requires an ethical commitment to doing research that actively and explicitly works to deconstruct structures and systems of oppression". For them, this involves "more than employing appropriate methodologies and following university ethical procedures" but also research practices and contributions as driving "ethical and social change". Thus, adopting such insights suggests a decolonial approach to landscape that necessitates both "attending to what and how we write" and the "commitment" to do our studies "differently" (2020: 642).

What does – or could – a different approach look like? Certainly, the application of collaborative or participant-led methods and forms of fieldwork might offer ways to intervene in and change the world for the better; destabilise hierarchies of knowledge, authority, and research design; and actively assist in realising a decolonial society (see Asher, 2019). Likewise, scholars have also documented ways of working in concert with or amidst marginalised and Indigenous people to support acts of resistance and liberation (Daigle and Ramirez, 2019). Such approaches are particularly important when considering the challenge of "how to nurture decolonising approaches while objecting to appropriation" (Barker and Pickerill, 2020: 643). To further illustrate such practices in application, we consider a well-established research partnership of non-Indigenous and Indigenous collaborators, and their ways of working in/with Indigenous knowledge and communities to articulate a means of studying what we might call "landscape" albeit on different terms.

> ***Pause for Thought:*** *Take a moment to consider your own positionality. How have you come to landscape studies in your work? How (if at all) have your working assumptions of what a landscape is been contested, extended, or complemented by the ideas discussed in this chapter so far?*

Doing Things Differently: Country

Arising from research begun in the late 2000s, and appearing in publication since 2012, the academic-Indigenous collective "Bawaka Country" demonstrates the potential for pursuing decolonial studies of landscape – or rather, *country* – in a different way. Here, we emphasise what their experience reveals about how landscape might be studied and represented.

Landscape as "Country": Bawaka Country names an "Indigenous-non-Indigenous, human and more-than-human research collective" (2015: 271): a collaborative group of scholars – both White and Indigenous-Aborigine – and Yolŋu community members and representatives, united by shared interest in the relationship between land and life in Bawaka – an area of North-East Arnhem Land, Northern Territory, Australia. The Aborigine-English term "country" – which incorporates "the land, the water, the animals, the plants, the rocks, the thoughts, the songs, the people and the ways they relate to each other" – understands land(scape) as an interconnected manifestation of space/place. Place is "vibrant and sentient"; creative and active in the (re)making of the world, rather than "a passive backdrop" (Bawaka Country et al, 2015: 270). Their work emphasises the importance of reimagining our ontologies of space/place and positions this as "central to the imperatives of decolonisation", in order that we might better "heed the call of place" and "understand the different ways of being in place" (Barker and Pickerill, 2020: 648).

In a subversion of academic norms and conventions, the land is figured as a "research partner", a (co-)author of research. In attributing "authorship" to the country – Bawaka – itself, the collective aims to practice adherence to "Rom" (Yolŋu law) that locates claims to authority and responsibility within a more-than-human community of which humans are only one part (see Bawaka Country et al, 2023). In this way, such work represents an effort to think landscape on terms outside of the Western-European dichotomies of nature/culture, being as several members of the collective do not recognise such a separation in their relationship with Bawaka, emphasising instead its "becoming" via relationships between humans and all manner of other beings (Bawaka Country et al, 2016).

Researching and Writing "Country": Thinking with country frames a *methodology* for doing research into this specific land(scape) that draws upon Indigenous practices and concepts, and thus demands a different way of *writing*. The collective often presents works as collaborative, poly-vocal accounts shuttling between academic non-Indigenous and Indigenous voices, with acknowledgement of which "we" is asserting itself in the dialogue (see Bawaka Country et al, 2019). Their work has also used the visual and textual figure of the "songspiral" to map and creatively represent Country as an intergenerational, communal entity, conveying something of its lived temporality (see Bawaka Country et al, 2022). This opposes the Western tendency to locate landscape in terms of a linear conception of history (past-present-future). As the collective writes:

> Spirits and memories of those who have died, for example, and who once dwelt in Bawaka continue as affective presences that constantly co-become and co-constitute the landscape. These lasting connections are made possible by the relationships which underpin them and by Bawaka itself. *Gurrutu* [the Yolŋu concept of kinship] suggests that place/space may be understood as more than living with(in) the physical landscape, it has a certain mobility; it is embodied and thus travels with the academic researchers as they return home (through their kinship relations and thus enduring emplacement within the Yolŋu landscape) and it reaches out to incorporate distant stars and space.
>
> (Bawaka Country et al, 2016: 465)

Refuting binaries: Celebrating an Indigenous ontology of "co-becoming", the work of the collective disrupts any sense that one can (or should) seek to disentangle the researcher from the researched. It is also an *active project* that works against (or despite) Western colonial norms structuring how academic research should be done, or how it should appear (Bawaka Country et al, 2023). Used as the basis for understanding land(scape), this way of thinking refutes any assumed separation of human actions from nonhuman forces and relations. Indeed, there can no longer be such a thing as "human" geography, for example (Bawaka Country et al, 2015: 279), within this schema. As the collective writes, their challenge to "colonial" perspectives on nature/culture is emphatic:

> Practising relational research through a methodology of attending means recognizing deeply that we live, write, work, eat and love as part of the world. We are not, can never be, distinct from it or from each other. We are related […] We need to recognize our connectedness, our becoming together. [...] With such an understanding methodologies based upon a premise of separation, of discrete human entities communicating to each other to a backdrop of passive nature become unthinkable, violent, unjust.
>
> (Bawaka Country et al., 2015: 279–280)

We have considered Bawaka Country here as a tangible example of doing landscape study with and through Indigenous ontologies in a way that neither appropriates nor "cherry picks". Importantly, this collective represents an *ongoing* work of engaging with specificity and difference in context; creatively, proactively, and with a view to dismantling lingering colonial assumptions about the world.

Decoloniality in the Heart of Empire

The tendency to locate Empire "elsewhere" can itself perpetuate a form of epistemological violence. As art historian Jill Casid (2005: xxi) writes of postcolonial studies, there is a need to "challenge the centre-periphery model through which the imperial metropole imagines itself at the centre of a world network and yet as separate and unchanged by exchange and contact with its colonies". As Radcliffe states, "although colonialism is often imagined to be 'over there', coloniality and decolonizing are present here and now" (Radcliffe, 2022: 6). Thus, as Chapter 6 considered the impacts of imperialism amidst European urban landscapes, so too might decolonial landscape studies attend to environs "seemingly far from the frontiers of settler colonialism" (Barker and Pickerill, 2020: 642).

An attention to the imbrication of memory, commemoration, and landscape can provide one such avenue. Importantly, memorialisation can be both monumental *and* mundane. In Dalnavert, a hamlet in the Scottish Highlands, there lies an unassuming monument to Sir John A Macdonald. Born in Scotland in 1815, Macdonald's family emigrated to Canada when he was five, where, in 1867, he became its first Prime Minister. The monument (a cairn) and plaque were unveiled "in his honour" (BBC News, 12 May 2010) (Figure 7.1). Fergus Ewing, the

Figure 7.1 Dalnavert memorial, Badenoch and Strathspey, Scotland, UK. Image by R. Hunt.

Government Minister sent to commemorate the occasion, celebrated Macdonald as "one of Scotland's most famous sons [who] had a huge impact on Canada and its people". This is a loaded statement: remembered by some for his nation-building policies, Macdonald has also been implicated in "cultural genocide" (The Truth and Reconciliation Commission of Canada, 2015) via the creation of residential schools (see de Leeuw, 2009) and enabling of famine conditions that forced First Nations populations onto reservations (Daschuk, 2013). The BBC coverage of the plaque's unveiling told nothing of this history, however. In Canada in 2020, Macdonald's statue in Montreal was one of the many taken down by protestors (and in this case, decapitated). His fate echoed that of slave-trader Edward Colston, whose statue in Bristol was dumped into the harbour during Black Lives Matter protests the same year. Symbols of what geographer Derek Gregory (2004) terms the "colonial present", they became a means of highlighting and challenging the economic, social, and political legacies of Empire as materialised in memorial landscapes.

Work in landscape studies examining the inclusions (or absences) of memorial landscapes offers one means to address the impact of colonialism in both the metropole and as traces left upon the (post) colonial state. As "authorised memory makers, instrumental in the creation of discourses of identity and belonging in public space" (Rofe, 2023: 633), monuments, memorials and statues ensure the ideology of Empire endures, "rendered in stone and steel", continuing to "present order and authority" or else perpetuate "subjugation and silence". In the same manner as the plaque at Dalnavert leaves a partial, neutralising account of history intact, other landscapes perpetuate colonial narratives in the present and require deconstruction. Part of decolonising landscape, then, as well as articulating its multifaceted nature, must also involve exposing its role in reproducing insidious, violent, and unjust discourses. Importantly, decoloniality in landscape studies should not only be tackled through examining the ways colonialism is remembered via landscape but also through the experiences of those diasporas whose lives and situation (within different landscapes) are shaped by (past) colonial regimes. We offer such an example in Box 7.3.

Box 7.3 Landscape, Citizenship and Heritage in Britain after Empire

Cultural geographer Divya Praful Tolia-Kelly (2010, 2016) has considered how supposedly quintessentially "English" landscapes in Britain can be connected to, or subverted by, a consideration of migrant and diasporic experience in the wake of Empire.

For example, she has considered how relationships between race, memory, and landscape infuse the subjectivities of South-Asian women living in contemporary London (Tolia-Kelly, 2016). The "landscape" – naming both a visual-discursive, idealised scene (of home or "Englishness"); *and* a set of lived, material surroundings – is central to the "ecology" of relationships by which citizenship, as a sense of identity, enfranchisement and belonging, takes shape for these women. Imagined and represented landscapes – displayed in hallways or recreated in gardens – are key sites for the articulation of diasporic lives, memories, and inheritances shaped partly under colonial rule. Tolia-Kelly's work thus emphasises "the multinational nature" of landscape as lived, sensory encounters with distant places offer "an orientation to particular environmental textures" (2016: 77) amidst British environments that both alienate and prompt nostalgia. Migrant experiences reveal landscapes as "multinodal" (2016: 68) or multiplicitous. Thus, "critical to identity formation in post-colonial terms", Tolia-Kelly argues that "remembered landscapes [...] situate historical narratives about the diasporic community" (2016: 10) and connect with everyday material culture and surroundings.

Elsewhere, Tolia-Kelly demonstrates how decolonial attention can trouble apparently "timeless" landscapes of Englishness. Specifically, heritage narratives prevailing within representations of Britain's "Roman Frontier" inform discourses of Englishness as associated with idealised notions of "Roman-ness", exceptionalism, civilisation, and longevity. Through research on Hadrian's Wall, Tolia-Kelly (2010) analyses how Britain's imperial project "shaped Roman history", prioritising "one narration of many that could be told" (Tolia-Kelly, 2010: 73–74). A conflation of Roman "civilisation" and imperial Britain (expressed via built landscapes– see Chapter 6) is predicated on associations between Romanness, Britishness, and *whiteness* that reflect the nineteenth-century heritage narratives that endure through the "frontier landscape" of Roman Britain.

However, such associations are destabilised by alternative histories and the "re-presencing of black figures in national history"; addressing absences that "compound historical mythologies and mistruths" (Tolia-Kelly, 2010: 78). For example, Tolia-Kelly elaborates how a history of Hadrian's Wall might also celebrate Severus Septimus (145–211 BCE), the "African Emperor", hailing from present-day Libya (Figure 7.2). Responsible for substantial (re-)building at the wall, Septimus' contribution, as significant as Hadrian's, has been historically underplayed. The result, of such omission (or

Figure 7.2 The Severan Tondo (c. 200 CE) classical panel depicting Severus Septimus (145–211 BCE), known as the "African Emperor", and who is believed to have been born in what is today Libya. Image © Jose Luiz Bernardes Ribeiro. Wikimedia, CC BY-SA 4.0.

"occlusion"), Tolia-Kelly argues, "is to smooth a notion of a cultural history of Britain where Roman-ness folds into Englishness and a sense of a nation built on a particular link to 'Europeanness' in terms of bodies, practices and moral landscape" (2010: 79). Moreover, challenging such depictions provokes resistance. In 2017, the BBC was criticised by some media commentators for allegedly "inaccurate" illustrations showing non-White Romans in Britain. The fact that archaeological and documentary sources evidence an ethnically-diverse Roman Britain is not enough to dismantle the idea of "Romanness" mobilised within racialised national politics, and such examples remind us that landscape "is itself a narrative" (Freiss and Jazeel, 2017: 15), mediating relations between people and environment in ways often exclusionary.

Decolonising Landscapes

This chapter has highlighted variously how landscape studies can critically engage and contribute to a decolonising agenda by unlearning certain assumptions, engaging alternative ontologies, and critically interrogating the perpetuation of colonial narratives via landscapes "at home". Much of this work represents a "metaphorical" decolonising: the dismantling of colonial through-structures, and the exploration of alternatives. In contrast, then, this final section considers more primarily decolonial forms of praxis: practical efforts at decolonisation as they impact the material relations between communities and the landscape. Indeed, as Tiffany Dang, echoing the sentiments of Tuck and Yang (2012), emphasises:

> Decolonization is most definitely not a metaphor, especially not in the context of landscape studies – a field primarily concerned with the study, design, and agency of land. If colonialism is about the control of land, then conversely decolonization requires the complete subversion of the power(s) controlling that land. Decolonization starts with land.
>
> (Dang, 2021: 1004–1005)

Consequently, we outline empirical examples of how decolonisation of the landscape might unfold via the (re)assertion of Indigenous involvement in the management or control over land and territory. If decolonisation names "not an 'and' [but] an elsewhere" (Tuck and Yang, 2012: 36), then these examples help to gesture towards such alternative landscapes and futures.

Decolonising River Landscapes

Amanda Thomas (2015) draws from her research into the management of riverine landscapes in Aotearoa-New Zealand – specifically the Hurunui river within the

Canterbury catchment – to elaborate how "Indigenous more-than-humanisms" might be given space in the context of planning decisions. To do so, she describes how members of the *iwi* (tribe) of the Ngai Tahu Māori were invited, as stakeholders, to join the Hurunui Waiau Zone Committee (HWZC) – a body created by the Canterbury Water Management Strategy to advise on and inform policy with regard to managing the river landscape – in response to growing regional tensions between settler and Indigenous communities.

Initially, Thomas describes Indigenous members of the committee encountering racism as their knowledge and insights are dispelled as "myth". The representatives became frustrated by dominant Eurocentric discourses of Enlightenment science and market economics, tending to "value rivers as a series of objects with enormous potential to produce economic growth" (2015: 979). River habitats were interchangeable, considered in terms of functions and services. Water merely allowed to flow, unimpeded, was a wasted resource when it could be harnessed for energy. By contrast, members of the *iwi* conceived the river in terms of a broader cosmology of *Whakapapa*: an understanding of human-environment relations as "the genealogical relationship between all things" (2015: 981), informing a sense of linked kinship. Thus, landscapes and their nonhuman inhabitants constituted "nonlinear family relations" (2015: 981) with whom humans had to exist, work, and live well on the land. Humans were not separate from river "Nature" because their involvement in river ecologies was a given, and required behaving with respect and care.

During the committee's deliberations, other river users found that Indigenous committee members, asserting a holistic ontology of the river's more-than-human ecology, provided a way of valuing non-economic engagements with the river landscape. As Thomas observes, in contrast to traditional Western "preservationist" conservation discourses, positioning the "wild" as a separate and distant place, members of the *iwi* saw human presence, use, and care for river ecologies as the basic premise for subsequent discussions about management. Thus, Māori ways of understanding the river landscape accommodated and gave voice to ways of engaging the environment for aesthetic or cultural purposes that were excluded by utilitarian economic frameworks. Such ways of using the landscape could then be included in planning discussions for the catchment, recast as "a complex knot of other things, constantly moving" (2015: 982) lived with and as part of. The introduction of an relational ethics approach, facilitated by the inclusion of the *iwi* on the river committee, led to both a more holistic approach to river management and an expansion of focus, including consideration of lagoons and the river mouth (contrasting previous focus on the upper regions), alongside proposals for increased controls on farming in the district to reduce the flow of nitrogen and phosphorous into the Hurunui river.

We include this example here because it demonstrates how the involvement of Indigenous groups in land management bodies can create opportunities whereby alternative ontologies of landscape are able to influence policy decisions. At the same time, representatives from the *iwi* are being asked to operate within a fundamentally colonial set of structures (river management relations defined by the settler state). Nevertheless, Thomas shows that non-Western ways of thinking can provide a vital route to articulating a more-than-human relational ethics, offering a means to forge negotiated partnerships between different stakeholders.

Seṅák̲w: Indigenous Lands in Colonial Worlds

Tiffany Dang offers another example, which examines the practicalities of a decolonised landscape: a $600 million residential development, sited beneath the Burrard Street Bridge in central Vancouver. Here, building has begun for 6,000 residential units across 11 towers, the tallest of which reaches 56 storeys. As Dang highlights, this development is "unique", both for its spatially economic design but also because it has been proposed by the Squamish Nation on a section of reservation land, "Kitsilano 6".

This development, named Seṅák̲w (Figure 7.3), disrupts the "settler status quo". It exists on an "urban" reserve: an area of land ascribed to the Squamish nation in 1868 and annexed by the city of Vancouver in 1913, displacing the resident population to the Howe Sound. This small, 10.5-acre fraction of the original reserve was returned to the Squamish nation after decades of legal wrangling and offered "a unique site for an urban residential project" owing to its jurisdiction under federal (rather than municipal) law, enabling a bypassing of planning regulation which would otherwise have restricted height, density, and car parking provision. The development could also address an affordable housing crisis through its provision of rental properties (70–90% of those proposed). Thus, Seṅák̲w "by virtue of its ownership and design [...] represents a disruption to the settler status quo [...] a small beacon in the pressing need for the dissolution of settler hegemony over urban landscapes" (2021: 1013).

Yet contradictions abound. Such reserve lands are embroiled within a colonial legacy that sees them unable to be owned by Indigenous nations. The land, designated to the Squamish nation for exclusive use by the Federal government of Canada, remains embroiled within a system of financial oppression through Section 89 of the Indian Act, which "restricts the ability for reserve land to be mortgaged or seized" with clear social, economic, and political consequences for the Indigenous livelihoods involved. The project, as an urban redevelopment, is also inherently a *capitalistic* venture. It therefore maintains the very system that is "inseparable from the structure of settler-colonialism which maintains settler-state power over Indigenous societies" (Dang, 2021: 1013). Ultimately, however, for Dang, "Seṅák̲w contributes to decolonization because it supports Squamish self-determination on Squamish land" (Dang, 2021: 1013). Following her example, landscape scholarship might further engage landscapes located "at the intersection of multiple ruptures" (2021: 1013) where practical forms of decoloniality are, imperfectly and not without contradiction, being carried out to unsettle forms of colonial control and produce new kinds of possibilities for Indigenous communities.

Figure 7.3 Seṅák̲w development in 2024, Vancouver, CA. Image by Geordie Milne and Kristopher Grunert. Wikimedia Commons, CC BY-SA 4.0.

Chapter Summary

Box 7.4 Key Concepts

Chapter 7

Postcolonialism
decolonise
decoloniality
Indigenous ontologies
country

Decolonising landscape studies is a complicated process of unsettling existing understandings, exploring new ways of working, and acknowledging pasts whilst opening onto alternative futures (Box 7.4). This is both a challenging task and an ongoing one. For Daigle and Ramirez (2019: 79), decolonial scholarship labels "a constellation in formation". As they elaborate:

> Constellations are in formation all around us, re-envisioning and re-embodying a politics of place by interweaving spatial practices of resistance, refusal and liberation. These historical and always emerging relationships across decolonial struggles transcend colonial boundaries by disclosing the interconnected terrain of racial capitalism, colonialism and white supremacy from one space to the next. More than this, through the spatial concept of constellations, differentially situated peoples are renewing and creating futures that have always been present in their/our own communities. These spatial formations of resistance and creation draw from the histories and geographies of Black, Brown and Indigenous peoples to re-root and re-route toward more accountable relations. And this what we see as the heart of decolonial geographies: that these stars form constellations to guide us toward decolonial futures.
>
> (2019: 82)

Fostering decolonial futures within landscape studies requires scholars to acknowledge the role that landscape studies – and associated institutions, individuals, and tools – have themselves played in the establishment, maintenance, and justification of colonial regimes. Thus, scholars must unlearn the very ways in which they work and consider the epistemological appropriateness of deploying existing approaches and concepts – even landscape itself – going forward. The difficulty is justified by the importance of this task: Landscape studies must be willing to accept their own inadequacies, and learn to work differently, work collaboratively, and work in ways yet unknown. This is, most certainly, an "ontological struggle of epic proportions which seeks to fundamentally shift how the world is known, who we are, what the world is and what we [as academics] do" (Barker and Pickerill, 2020: 643).

It is important that colonialism and its violent effects are not assumed to exist somewhere else. The implications of coloniality are persistent and pervasive both "here" and "there", whatever your geographical location. The decolonial turn is fundamentally interdisciplinary and demands attention to our topics, sites, and tools but also to our ethics, practices, and interpersonal relations (Radcliffe 2022). Crucially, as we have discussed, scholars of landscape must "study up": engage with this history of their subject and critically engage with the dominant presumptions which have shaped landscape studies. It is through this learning, and unlearning, that we enable ourselves to explore the way that decoloniality "undoes, disobeys, and delinks" from colonial logics and legacies, "constructing paths and praxis toward an otherwise of thinking, sensing, believing, doing and living" (Mignolo and Walsh, 2018: 4). Ultimately, students of landscape studies, no matter where their interests lie, need to be aware of the ways in which landscape can be decolonised, done differently, unlearned, or even rejected, as the circumstances and subjects of our work demand.

Suggested Seminar Activities

Activity 1: Review Chapter 8 – "Twenty-five Indigenous Projects" in *Decolonizing Methodologies (3rd Ed.)* by Linda Tuhiwai Smith (2021). Identify one of the twenty-five project areas/themes she identifies and, in a group, consider how this project focus might be realised in the context of studying a specific landscape in a settler-colonial context.

- *The group should aim to produce a short research project proposal to present to the class.* This project should outline how a programme of decolonial landscape research might be undertaken in this landscape, and the potential outcomes for affected communities and stakeholders that such a project might realise.

Activity 2: Identify an example of a notable, yet contentious, historical monument or memorial situated within the boundaries of a former imperial state. Research the history that the monument is designed to celebrate and the existing debates/tensions around its existence today to inform understanding of why such a structure might (a) reproduce colonial structures in the present; (b) be targeted in the context of "decolonial" practice.

- *As a group, devise a creative way of intervening in/at this monument to disrupt/subvert its intended reading and unsettle or reimagine its role in the landscape.* Consider, for example, an artistic, activist, or performative intervention that could promote the need to think (differently) about this object and the role of Empire in continuing to affect our experiences of landscape.

Recommended Reading

https://bawakacollective.com/ – The homepage of the Bawaka Collective: an Indigenous and non-Indigenous, human-more-than-human research collective advancing decoloniality via their creative, collaborative and community-focused approach to land.

Dang, T. 2021. Decolonising landscape. *Landscape Research* 46: 1004–1016.

Tolia-Kelly, D. 2010. Narrating the postcolonial landscape: Archaeologies of race at Hadrian's Wall. *Transactions of the Institute of British Geographers* 36(1): 71–88.

References

Asher, K. 2019. Fieldwork. In: *Antipode* Editorial Collective. Eds. *Keywords in Radical Geography: Antipode at 50.* Oxford: Wiley-Blackwell, pp. 123–127.

Barker, A and Pickerill, J. 2020. Doings with the land and sea: Decolonising geographies, Indigeneity, and enacting place-agency. *Progress in Human Geography* 44: 640–662.

Bawaka Country et al. 2015. Working with and learning from country: Decentring human author-ity. *Cultural Geographies* 22(2): 269–283.

Bawaka Country et al. 2016. Co-becoming Bawaka: Towards a relational understanding of place/space. *Progress in Human Geography* 40(4): 455–475.

Bawaka Country et al. 2019. Goŋ Gurtha: Enacting response-abilities as situated co-becoming. *Environment and Planning D: Society and Space* 37(4): 682–702.

Bawaka Country et al. 2022. Songspirals bring country into existence: Singing more-than-human and relational creativity. *Qualitative Inquiry* 28(5): 435–447.

Bawaka Country et al. 2023. Author-ity of/as Bawaka Country. *Australian Archaeology* 89(1): 75–77.

BBC News. 12 May 2010. Life of Sir John A. Macdonald honoured in Dalnavert. http://news.bbc.co.uk/1/hi/scotland/highlands_and_islands/8675830.stm

Bledsoe, A. and Jamaal Wright, W. 2019. The pluralities of Black geographies. *Antipode* 51(2): 419–437.

Bradley, J. and Kearney, A. 2018 'You whitefellas pull it all apart': Epistemic learnings in exploring landscape. In: Howard, P., Thompson, I., Waterton, E., and Atha, M. Eds. *The Routledge Companion to Landscape Studies.* London: Routledge, pp. 288–298.

Britton, C. 2014. *Language and Literary Form in French Caribbean Writing*. Liverpool: Liverpool University Press.

Casid, J. 2005. *Sowing Empire: Landscape and Colonizati*on. Minneapolis: University of Minnesota Press.

Connell, R. 2007. Southern Theory: the Global Dynamics of Knowledge in Social Science. Allen and Unwin: Sydney.

Daigle, M. and Ramirez, M.M. 2019. Decolonial geographies. In: *Antipode* Editorial Collective. Eds. *Keywords in Radical Geography: Antipode at 50.* Oxford: Wiley-Blackwell, pp. 78–84.

Dang, T. 2021. Decolonising landscape. *Landscape Research* 46: 1004–1016.

Daschuk, J. 2013. *Clearing the Plains: Disease, Politics of Starvation, and the Loss of Aboriginal Life*. Regina: University of Regina Press.

de Leeuw, S. 2009. 'If anything is to be done with the Indian, we must catch him very young': Colonial constructions of Aboriginal children and the geographies of Indian residential schooling in British Columbia, Canada. *Children's Geographies* 7(2): 123–140.

Derickson, K.D. 2017. Urban geography II: Urban geography in the age of Ferguson. *Progress in Human Geography* 41(2): 230–244.

Driver, F. 2001. Geography Militant: cultures of exploration and empire. Maiden Mass, USA: Blackwell.

Friess, D. and Jazeel, T. 2017. Unlearning "Landscape". *Annals of the American Association of Geographers* 107: 14–21.

Gregory, D. 2004. *The Colonial Present.* Oxford: Blackwell.

Harley, J.B. 2001. *The New Nature of Maps: Essays in the History of Cartography.* Baltimore: John Hopkins University Press.

Hawthorne, C. 2019. Black matters are spatial matters: Black geographies for the twenty-first century. *Geography Compass* 13(11): e12468.

Jacobs, J.M. 1996. *Edge of Empire: Postcolonialism and the City.* London: Routledge.

Jazeel, T. 2013. Dissimulated landscapes: Postcolonial method and the politics of the space in southern Sri Lanka. *Environment and Planning S: Society and Space* 31: 61–79.

Jazeel, T. 2019. *Postcolonialism.* Abingdon: Routledge.

Lorde, A. 2019. *Sister Outsider*. London: Penguin Classics.

McDowell, L. 1992. Doing gender: Feminism, feminists and research methods in human geography. *Transactions of the Institute of British Geographers* 17: 399–416.

McKittrick, K. 2006. *Demonic Grounds: Black Women and the Cartographies of Struggle.* Minneapolis, MN: University of Minnesota Press.

McKittrick, K. 2011. On plantations, prisons, and a black sense of place. *Social and Cultural Geography* 12: 947–963.

Mignolo, W. and Walsh, C. 2018. *On Decoloniality: Concepts, Analytics, Praxis*. Durham, NC: Duke University Press.

Miller, K. 2014. *The Cartographer Tries to Map a Way to Zion.* Manchester: Carcanet.

Mohanty, C. 1984. Under Western eyes: Feminist scholarship and colonial discourse. *Boundary* 2(12): 333–358.

Moulton, A.A. 2023. Towards the arboreal side-effects of marronage: Black geographies and ecologies of the Jamaican forest. *Environment and Planning E: Nature and Space* 6(1): 3–23.

Nelson, C. (2016). *Slavery, Geography and Empire in Nineteenth-Century Marine Landscapes of Montreal and Jamaica.* New York: Routledge.

Noxolo, P. 2017. Introduction: Decolonising geographical knowledge in a colonised and re-colonising postcolonial world. *Area* 49(3): 317–319.

Noxolo, P. 2022. Geographies of race and ethnicity 1: Black geographies. *Progress in Human Geography* 46(5): 1232–1240.

Pugh, J. & Chandler, D. 2023. The world as abyss: The Caribbean in critical thought and the Anthropocene. London, UK: Westminster University Press.

Radcliffe, S. 2022. *Decolonising Geography.* Cambridge: Polity Press.

Rofe, M. 2023. Memorial landscapes, recognition, and marginalisation: A critical assessment of Adelaide's 'cultural heart'. *Landscape Research* 48(5): 632–646.

Said, E. 1978. *Orientalism.* London: Penguin.

Spivak, G. 2004. Can the Subaltern speak? In: Cain, P. and Harrison, M. Eds. *Imperialism: Critical Concepts in Historical Studies*, vol. 3. London: Routledge, pp. 171–219.

Thomas, A. 2015. Indigenous more-than-humanisms: Relational ethics with the Hurunui River in Aotearoa New Zealand. *Social and Cultural Geography* 16(8): 974–990.

Tolia-Kelly, D.P. 2010. Narrating the postcolonial landscape: Archaeologies of race at Hadrian's Wall. *Transactions of the Institute of British Geographers* 36(1): 71–88.

Tolia-Kelly, D.P. 2016. *Landscape, Race and Memory: Material Ecologies of Citizenship.* Abingdon: Routledge.

Truth and Reconciliation Commission of Canada. 2015. Honouring the truth, reconciling for the future. Available via: https://web.archive.org/web/20200430162813/www.trc.ca/assets/pdf/Honouring_the_Truth_Reconciling_for_the_Future_July_23_2015.pdf

Tuck, E. and Yang, K.W. 2012. Decolonization is not a metaphor. *Decolonization: Indigeneity, Education and Society* 1(1): 1–40.

Tuhiwai Smith, L. 2021. *Decolonizing Methodologies: Research and Indigenous Peoples.* 3rd Ed. London: Zed Books.

8 Bodies in the Landscape

Introduction: Cobblestones

Designed landscapes, particularly those of large English country estates, exemplify a visual approach to landscape, as discussed elsewhere (see Chapter 2). Their form reveals an attempt to put a vision of "nature" on display as befits ideological positions concerning humans' relationships to their environment, and the organisation of society. Materialising a particular way of seeing the world, bound up with privilege, money, and landownership, these landscapes exude the values of the powerful. And yet, like any landscape, visiting such places produces a different experience for the visitor depending on the characteristics of their specific body. Sloping lawns, open vistas, or box-cut hedges are apprehended through the senses and physiological capacities of each individual, never in quite the same way.

As part of an online seminar in 2021, entitled "Whose Landscape", the environmental and medical historian Clare Hickman specifically addressed the ways in which estate landscapes afford particular conditions for bodies to interact with the environment, as well as persistent forms of bodily exclusion. Whilst many once-private country estates are now open to the public, their material form continues to limit those bodies that move easily through them. Take, for example, a cobblestone path. Hardly a surface affording mobility for all, cobblestone transforms the landscape into something bumpy, jarring, and difficult to walk, wheel, or be wheeled across. In such instances, it becomes clear that landscapes are *felt* as well as apprehended through vision. Moreover, attention to these encounters with stone highlights how such estate gardens were designed to be consumed by *able* bodies. Interests in "embodiment" of the kind that Hickman sets out therefore demonstrate the potential value for studies of cultural landscape which go beyond representations and consider direct, corporeal forms of knowing the world.

The philosophical approach and methods of phenomenology represent one way that scholars of landscape have sought to recover and attend to bodily experience. This approach begins from the premise that all perceiving subjects are *bodies in the world*. Thus, the interpretation of landscape becomes a practical question of bodily experience: how is it felt, smelt, heard, moved across or through, *as well as* seen? Such an approach offers one means of addressing the concern emerging in the latter 1990s that too much emphasis in cultural studies had been placed on

DOI: 10.4324/9781003172451-9

representation (images, texts, discourses, or categories of identity) and materiality (in the sense of a Marxist focus on historical materialism and political ecology) at the cost of how the world was, first and foremost, engaged with by living beings. Indeed, where previous approaches placed distance between the self and landscape, phenomenological approaches argue such distance is impossible. Whilst there are well-founded anxieties and critiques accompanying phenomenology (and its use in landscape study – see Wylie, 2018) concerning its overly abstract, imprecise, or (to some) challenging ideas, it has been, in some form, a feature of landscape scholarship since the 1960s (if not before). Such longstanding presence reflects, perhaps, the general sense that lived experience of landscape does matter for its conceptualisation. This chapter thus aims to introduce and outline the ways in which phenomenological ideas, particularly around the notion of "embodiment", offer a valuable way of understanding cultural landscapes.

Chapter Outline

The chapter introduces the philosophical approach of "phenomenology", establishing its origins and key contentions, before setting out its contribution for the study of cultural landscapes in the context of a broader interest in emotions, feelings, sensations, and relationships developed elsewhere in the textbook (see Chapters 9–12). In phenomenology, knowledge of the environment is framed as fundamentally embodied as the experiences comprising such knowledge unfold through or with the body, and its capabilities. The introductory discussion of phenomenology offered here focuses upon how we might understand the body's role in producing knowledge of the environment, and the ways in which this has been conceptualised in different ways since the 1970s. The chapter also considers a range of critiques – or anxieties – associated with a phenomenological framing of landscape, including the way in which "subjects" are presented; its persisting romanticism; and its (for some, limited) potential to offer economic and political critique. In looking towards how phenomenology has been expanded and developed in recent years, the chapter sets out some ways in which these critiques have been addressed in the realm of "post-phenomenology" and its interest in bodily difference.

Knowledge and the Body

For phenomenology, the body is central to the production and experience of cultural landscapes because it is through the body that we perceive and "know" our world. The phenomenological approach offers a direct critique of the epistemological approach (or basic assumptions concerning what knowledge is and how it is made) explored within previous chapters (e.g. Chapters 2, 4, and 5). That is, that the human mind apprehends and makes sense of the world in ways which tend to forget or leave out the body. Understanding the fundamental tenets of phenomenology therefore means briefly reflecting on some basic philosophical questions concerning knowledge itself; how it is produced, structured, evaluated, and with what effects.

"Embodied" Knowledge

As discussed in previous chapters, the sense abounds of landscapes as objects to be identified and scrutinised by an expert eye; documenting, describing, decoding, or explaining their character and cultural significance. Landscape knowledge is accumulated through careful, distanced *observations.* The body or bodily state of the researcher is of negligible importance: it can be removed (discounted or, by way of training, disciplined away) from any resulting account. This approach to knowledge production implies what John Wylie terms a "spectatorial epistemology" (2007: 145), presuming "an external pre-given reality observed and represented from a detached position by an independent perceiving human being". In this way, "seeing and knowing are conflated" (Rose 1993: 86) as the specific social characteristics (in Chapter 2) or bodily capacities of the "seer" are exorcised from this activity.

A "spectatorial" version of knowledge hinges on an underlying binary in the Western philosophy of thought between, on the one hand, the mind (and thinking) and, on the other, the body (and sensation). This binary can also be expressed in terms of human society or culture as set apart from nature or the nonhuman. Edmund Husserl, often considered the founder of modern phenomenology, terms this outlook the "natural attitude": the taken-for-granted sense that the world "out there" is separate from society or thought and exists as the object for our actions or assertions (see Husserl, 1999). The writings of seventeenth-century philosopher Rene Descartes are often seen as the basis for this attitude, with Descartes famously arguing that the soul and spirit of any (human) being existed within the mind and thus one's sense of selfhood began with the awareness of one's capacity for thought ("I think therefore I am").

In turn, if the mind (rather than the body) was the essence of being and source of rational knowledge, then the production of knowledge (for "science") required removing the body, its feelings, and instincts, from the equation in order to view the world dispassionately and unmediated by distraction or sentiment. Thus, the Cartesian dualism (the separation of mind and body) informs our held idea of "objective knowledge" within which detachment and distance are necessary standards for "truth". As spatial theorist Edward Relph highlights, this mode of thinking "provided a basis and justification for science that involved no appeals to animating souls or spiritual forces, but required solely the power and certainty of human thought" (2016 [1980]: 28). The body and its senses (as well as the material world) are confined to a lesser realm.

Contrary to the Cartesian worldview, Husserl and later thinkers such as French philosopher Maurice Merleau-Ponty would conceptualise knowledge as *embodied.* That is, inseparable from, and located in, the experience and existence of the body. Merleau-Ponty would significantly challenge the notion of an "objective" Cartesian mind-subject in an influential book, *Phenomenology of Perception* (1962). For him, knowledge was "felt" and produced by a subject that was also always a body: a body-subject. To be sentient, then, meant not only thinking but feeling. Indeed, these were hardly exclusive conditions. Rather, each *necessitated* the other: to feel is to think, to think is to feel. As Merleau-Ponty would propose, one should not presume the world to comprise some separate domain awaiting our understanding or explanation. Instead, he argues, the world "is the natural setting of, and field for, all my thoughts and all my explicit perceptions" (Merleau-Ponty, 1962: xi). Phenomenology, so defined, was a "style of thinking" to describe (rather than explain) "the essence of perception"; or how the subject made sense of their world as it was being "directly experienced" (Merleau-Ponty, 1962: vii–viii).

This is a big shift in how we are understanding our relationship with the world. As anthropologist Tim Ingold (2015: 84) writes, knowledge considered thus involves the body in all its capacities: "it is because we can see that we experience light, because we can hear that we experience sound, and because we can touch that we experience feeling". The body is not ancillary or additional to the mind, a container that gets in the way. To think *is* to be embodied: the dichotomy of mind/body is dispensed with. Embodiment does not denote a particular *kind* of experience, but the very condition of experience: all experience of the cultural landscape "is" embodied, because we are bodies having those experiences of an environment, and perceiving it as a landscape. The world is made knowable as a set of surroundings that the body "holds" in relation to itself as it moves, senses, and feels.

Contextualising Interest in Bodies and Landscape

In upending the persistent binaries of knowledge/feeling, body/mind, phenomenology presents a major challenge to the way that many conceive of subjectivity, knowledge, and perception. From a phenomenological basis, the cultural landscape becomes something *inhabited* rather than merely observed, interpreted, or read. The body, and bodily experience, is cast as the starting point from which to conceptualise, understand, and research the landscape. Such attention to "corporeality" (the state of being or having and experiencing through a physical body) and bodily practice (the very *doing* of things) has a longer lineage in the context of landscape writing, with important precedents found across various writings about landscape from the 1950s; as well as the philosophy noted above of Husserl, Merleau-Ponty, and other individuals such as Martin Heidegger produced during the twentieth century. Indeed, even in Carl Sauer's elaboration of cultural landscape study, he makes explicit that knowledge of the environment requires direct, physical contact (he even uses the term "phenomenology" himself) in the form of in-situ fieldwork. Therefore, we briefly consider some figures within those earlier writings here.

The work of American writer J.B. Jackson is often figured as a key reference point for more recent attention towards the everyday, practical cultural landscape (see Cresswell, 2003). Interested in landscape design, and with a background in architecture and planning, Jackson's early work on landscape saw him draw upon aerial photography. He soon turned his attention to more everyday American landscapes (see below) collating his thoughts (and those of others) in the journal-magazine *Landscape* (1951–1999), which he founded and edited until 1968.

Across a diverse output from the 1950s and 1960s, Jackson emphasised that the orthodox view of landscape as something visually surveyed failed to capture its lived-in qualities. His cultural landscape is named "a composition of man-made [sic.] or man-modified spaces to serve as infrastructure or background for our collective existence" (Jackson, 1984: 8). Humans were neither "spectators" nor landscapes "a work of art"; rather they comprised the "temporary product of much sweat and hardship and earnest thought" (Jackson, 1997: 343). The environments that interested him were inhabited and produced through everyday life. Landscape study concerned depiction of a material or "workaday" world populated with houses, roads, and sidewalks. As Wylie writes, this was "a world on the move and in the making rather than fixed or framed" comprising ordinary, lived-in surroundings wherein inhabitants were "improvising and elaborating a life far removed from both metropolitan power and abstract intellectual theorisation" (Wylie, 2007: 43). Thus, Jackson's interest in "vernacular landscape" turned away from documenting a static morphology in favour of actual "experience" (Cresswell, 2003: 217).

Jackson's assertions that cultural landscapes were lived in as well as observed are echoed in the work of "humanistic" geographers including Anne Buttimer, David Seamon, Edward Relph, and Yi-Fu Tuan. In social theory, the French philosopher Michel de Certeau would elaborate a similar framework for the study of everyday social life in his 1980 work *The Study of Everyday Life*, translated into English in 1984. De Certeau (1984) proposed urban environments be examined in terms of the practical, often tactical ways in which they were engaged with at ground level, often subverting or contrasting how such landscapes were planned for or viewed out "from above". Thus, these works understood the subject and their surroundings in ways that foregrounded the individual, fleshy body and its involvement in a tangible world. The cultural landscape encapsulated "everything I see and sense when I'm outdoors"; incorporating visual stimuli but also "the smell of gasoline fumes, the feel of the wind, and remembered experiences" (Relph, 2016 [1981]: 22).

Another important touchstone, or rather comparative, for Western landscape phenomenology is the many Indigenous understandings of landscape and nature that have long given credence to sensory forms of knowing (Nash, 1996 – see Chapter 7). One must be doubly cautious here. Firstly, too often the tendency to locate non-Western ontologies into the past within a chronology of knowledge, or to equate or conflate diverse ways of knowing with specific histories and contexts, or to (even inadvertently) position contemporary Western theoretical approaches as in some way more advanced or "developed" than others, simply reproduces the racist logic of colonialism. Secondly, the romantic tendencies of phenomenology (see below) risk evoking the similarly colonial sensibility that non-Western societies are somehow *more* embodied or connected to the environment than Western ones (see Jacobs, 1994). As discussed above, phenomenology asserts that *all* knowledge is "embodied", whatever its context, rather than implying some knowledge (or cultures) are *more* embodied than others – a common slippage. The key point here is that there are many examples of knowledge frameworks that take account of – even begin from – the body's central role as the conduit for environmental experience that sit outside of the "lineage" of phenomenology.

Across the accounts of landscape offered across this textbook, the body of the subject is generally present *in some way*. What differs is perhaps the extent to which bodily capacities of different kinds are explicitly discussed, or given prominence, in how cultural landscape is defined or analysed. As we now explore, the promise of phenomenology (and comparable approaches) is their expanding sense of landscape from the screen onto which cultural values are projected into a living, vibrant field of matter and relations (see Chapter 10) within which we are immersed and through which we become aware of the fact of *being alive* (Ingold, 2011).

Embodied Landscapes

Understanding bodies as central to the process of knowledge production leads to understanding cultural landscapes as "doings": that is, as environments that are performed or take shape by way of various practices. This focus upon doing and practice informs work attentive to how everyday actions shape, and are shaped by, landscape. The cultural landscape becomes a relational entity: it is something produced through relationships with the world, rather than a category or understanding project onto the world. It is unfixed, contingent, and continuously in-formation via the tensions or meetings that exist between bodies and matter.

Whilst early humanistic attention to landscape became overshadowed by interest in cultural values, identity, and discourse through the 1980s and early 1990s, interest in the landscape as "embodied" has seen resurgence over the past two to three decades. The application of ideas from phenomenology has been observed across the humanities in fields as diverse as archaeology, cultural anthropology, architectural theory, performance studies, and geography. Collectively, whilst admitting great diversity, this work advances a sense of landscape seen as "more connected to the forces that shape our lives" (Cresswell, 2003: 277).

A reengagement with phenomenological ideas and themes comprises something of the broader increased interest in practice, material, and affective dimensions of culture and social life observable across humanities and social science disciplines during this period, producing accounts placing renewed emphasis on action, movement, sensation, and perception. A key influence is the collected approaches and ideas often grouped under the heading of "non-representational theory" (see Thrift, 2008). Work in this vein critiques a constructivist epistemology. That is, the idea that the ways in which the world is organised, given meaning, categorised (e.g. in terms of identity or type) and represented are formed or "constructed" in the mind and then applied onto a pre-existing, material world. Such an account, it has been argued, however, renders social life inert – or "dead" – because its content is simply held to be the expression of predetermined, agreed social conventions or discourses. It leaves no (or too little) space for the emergent, surprising, and, crucially, non-cognitive (e.g. unconscious or pre-conscious) or non-linguistic forms (e.g. outside of language) ways we engagement with landscape (see Thrift and Dewsbury, 2000).

Non-representational theory has, as geographer Hayden Lorimer (2005: 83) suggests, "become as an umbrella term for diverse work that seeks better to cope with our self-evidently more-than-human, more-than-textual, multisensual worlds". It is important to note that non-representational work does not mean such scholarship is *anti*-representational. Rather it is "more-than-representational" in its accounts of the world (Lorimer, 2005: 84), recognising that representational practices *as well as* bodily practice, feeling, and relation are all implicated in "how life takes shape" (Lorimer, 2005: 84).Thus, the approaches discussed in this chapter (and Chapters 9–11) represent various attempts to foreground the body and expand a sense of how the environment is known outside of cultural codes or categories, as something felt, sensed, and experienced.

> ***Pause for thought:*** *In what ways is the landscape encountered through the body (beyond the visual)? Discuss, or list, all the ways the landscape is perceived or "known" by your body (e.g. senses).*

Landscape Phenomenologies

How, given all this intellectual context, does landscape phenomenology *get done*? An example of an early, and influential, treatment of landscape in these terms concerns anthropologist Tim Ingold's conception of the environment's "temporality". In an influential essay on the topic, Ingold proposes understanding landscape and its formation in relation to bodies, their involvement, and experience through time. In doing so, Ingold also critiques the idea of landscape as a visual construct or "cultural image". Rejecting Daniels and Cosgrove's (1988: 1 – see Chapter 2) definition of landscape as "a cultural image, a pictorial way of representing or symbolising surroundings", and its "division between inner and outer worlds – respectively of mind and matter, meaning and substance – upon which such a distinction rests" (Ingold, 2000: 191), he outlines an alternative:

> The landscape, I hold, is not a picture in the imagination, surveyed by the mind's eye, nor however is it an alien and formless substrate awaiting the imposition of human order. [...] As the familiar domain of our dwelling, it is *with* us, not *against* us, but it is no less real for that. And through living in it, the landscape becomes a part of us, just as we are a part of it.
>
> (Ingold, 2000: 191 – orig. emph.)

For Ingold, the landscape comprises the broader environment with which one is engaged (in doing, or being). Accounts privileging the representation of the environment (in art, in language, in thought) presuppose a false division between the symbolic (cultural ideas imposed *upon* something) and the material (some "thing" awaiting meaning to be bestowed). This division, for Ingold, pervades much cultural scholarship and casts landscape as either "a neutral external backdrop to human activities" or else "a particular cognitive or symbolic ordering of space" (Ingold 2011 [2000]: 189). Naming this view of the world as a blank slate to be written over as the "building perspective", he charges that it conceives of humans as operating within "worlds [which] are made before they are lived in" (2011 [2000]: 179). The humans (or culture) and environments (or nature) are enshrined as separate spheres of existence. Within the building perspective, "meaning [...] is laid over the world by the mind" rather than emerging from "the relational contexts of the perceiver's involvement in the world" (Ingold 2011 [2000]: 51).

An alternative stance, which Ingold names the "dwelling perspective", provides a phenomenological counter. The concept of "dwelling" has a long history in the humanities, drawing especially from the output of German philosopher Martin Heidegger, as it articulates the ways in which worlds are made through the course of their being "brought to hand" by a living subject. In this vein, for Ingold, the dwelling perspective proposes that, "through being inhabited [...] the world becomes a meaningful environment" (2011 [2000]: 173). Drawing upon Heidegger, Ingold describes dwelling in terms of the practical activities of both humans and other creatures, engaged with their environments and the objects that comprise them. Via their bodies, and specific bodily capacities, different materials or things afford different possibilities for action, sensation, and thus meaning. Emphasising the "agent-in-its-environment" (2011 [2000]: 173) and its ongoing involvement in a living world, the dwelling perspective renounces the dualism of body and mind. The individual in the landscape is "a being immersed from the start, like other creatures, in an active, practical and perceptual engagement with the constituents of the dwelt-in world" (Ingold, 2011 [2000]: 42).

Ingold's "ontology of dwelling" informs his proposed replacement of landscape with "taskscape". This term, he suggests, better captures the durational (through time), rhythmic, and active nature of cultural landscapes as embodied. What a landscape "means" reflects people's changing interaction with the world over time. Dwelling is always happening, but *how* it happens changes: as we move through the world, accumulate a sense of its character or what our bodies can do within it, and as it is transformed through its interactions with beings and forces of all kinds. Indeed, landscapes – or taskscapes – can be considered continuously *performed*. Performance, here, refers to the "doings" or actions, which create place. As the term taskscape seeks to imply, "landscape is the world as it is known to those who dwell therein, who inhabit its places and journey along the paths connecting them" (2011 [2000]: 193). Cultural landscapes, as embodied taskscapes, take shape via "the everyday project of dwelling in the world" (2011 [2000]: 191). This also does not mean that representations of landscape are unimportant, but that we might consider them in a different way, for what they reveal about the phenomenology of landscape (see Box 8.1). Not "a totality" that can be looked upon or grasped from some detached position, landscapes are the very *contexts* for and in which looking, thinking, feeling, and knowing are made possible.

Box 8.1 Landscape Paintings as Records of Embodiment

A phenomenological approach to landscape challenges the way representational texts (such as paintings) are understood in landscape study. A constructivist approach to landscape and representation (see Chapter 2) might examine how conventions of perspective, objectivity, and aesthetics inform our perception and evaluation of the environment, as well as reproduce specific ideologies (such as concern for landownership). Alternatively, landscape scholars have utilised painted imagery to express its *embodied* characteristics (e.g. Relph 2016 [1981]; Ingold 2011 [2000]) as representations might also be used to prompt consideration of how environments are (or were) experienced.

Both Tim Ingold and Edward Relph use Pieter Bruegel's *The Harvesters* (1565) (Figure 8.1) to offer a more sensuous account of landscape. For example, Ingold asks the reader not to look for themes or symbolic representations within the work but to reflect on the world it captures and imagine the scene as one amidst which they might "dwell":

> Imagine yourself, then, set down in the very landscape depicted, on a sultry August day in 1565. Standing a little way off to the right of the group, beneath the tree, you are witness to the scene unfolding about you. And of course you hear it too, for the scene does not unfold in silence. So used are we to thinking of the landscape as a picture that we can look at, like a plate in a book or an image on a screen, that it is perhaps necessary to remind you that exchanging the painting for "real life" is not simple a matter of increasing the scale. What is involved is a fundamental difference of orientation. In the landscape of our dwelling, we *look around.*
>
> (Ingold, 2000 [2011]: 202 – orig. emph.)

Taking us through elements of the scene, Ingold invites understanding of a landscape perpetually in motion. Considering the hills and valleys, Ingold argues we tend to categorise such features in ways that are socialised and conventional (as picturesque or rugged, for instance). In contrast, he suggests hills and valleys as known via bodily movement, the experience of rise and fall. Their contours are *felt* (when walking and the muscular effort entailed) as much as seen. Even if standing still, these features are seen in widths evoking movement as the eye travels up or down a gradient. Moreover, their material form is *made* through movement; whether over generations of feet wearing a track or geological timescales exceeding normal human perception. The river does not flow between pre-cut banks; *it cuts the banks by flowing*. Thus, a landscape is made via movement and time; it is continually happening and being formed. Such realisation is prompted by *The Harvesters*.

Figure 8.1 *The Harvesters* (1565). By Pieter Bruegel the Elder. Metropolitan Museum of Art/Wikimedia Commons, Public Domain.

Accounting for the people in Bruegel's scene, Ingold highlights how one not only looks at, or feels, but *hears* a taskscape and "the air […] full of sounds of one kind and another":

> Though the folk beneath the tree are too busy eating to talk, you hear the clatter of wooden spoons on bowls, the slurp of the drinker, and the loud snores of the member of the party who is outstretches in sleep. Further off, you hear the swish of scythes against the cornstalks and the calls of the birds as they swoop low over the field in search of prey. Far off in the distance, wafted on the light wind, can be heard the sounds of people conversing and playing on a green, behind which, on the other side of the stream, lies a cluster of cottages.
>
> (Ingold, 2000 [2011]: 207)

Again, this treatment of *The Harvesters* undoes representational accounts of landscape to replace them with a sense of a world that is living, practical, and durational. Whilst Ingold's approach bypasses the social and historical context of the image's production and interpretation (never mind questions of power or inequality manifest in the scene depicted), his embodied approach to the image demonstrates how landscape representations, subject to creative or expansive readings, can be used to make tangible the experiential, material, and sensory dimensions of the lived environment.

Pause for thought: *Choose an image of a landscape. Consider all the ways in which your bodily senses could come into play if you were to encounter that scene in person. How might this change your perception of this environment? How might your sensory experience of that environment change with the passage of time?*

Ingold's work is typical of the resurgence of phenomenological, embodied, more-than-representational attention to landscape that gained traction during the 1990s (see MacPherson, 2010). Whilst visual understandings of landscape remain important, this is as part of a broader consideration of how bodies engage the environment. Indeed, acts of seeing landscape are hardly *disembodied* acts (see Chapter 2). Experiences of visual perception can change as other sensory stimuli (like the pain of blisters, or bodily exhaustion) envelope and infuse our judgement of our surroundings (Wylie, 2005).

Today, a wealth of practice-focused work in landscape studies is often collected under the umbrella of "landscape phenomenologies" (Wylie, 2018), accounting for "embodied acts of landscape" (Lorimer, 2005: 85). Recent examples include Heidi Mäkelä et al's (2024) exploration of the practices of forest yoga in Finland to examine how embodied understandings of space intersect with national imaginaries of nature. In the Lake District, Sarah Nettleton (2015) explores the embodied experience of fell-running to consider how runners know the landscape in ways challenging conventional, picturesque framings of the National Park. In the context of contestations over territorial borders in the Middle East, Olivia Mason (2021) draws from her own experiences of trail walking in Jordan to ground such debates in direct bodily encounters with landscape. Understanding the world through a foregrounding of the body has also found influence beyond the academy, to influence landscape policy (Box 8.2).

Box 8.2 Phenomenological Approaches to Conservation Practice and Land Management

Relational, practical, and phenomenological ideas of landscape have been taken forth in conservation debates in Aotearoa New Zealand by the conservationist Mick Abbott. Influenced by the work of Tim Ingold, Abbott argues that "Ingold's position [regards landscape] demands a different approach" (2008: 154) to conservation practice and land management. He proposes that a sense of bodies, and the subsequent relational experience of landscape, is brought into the framing of management decisions particularly concerning so-called wilderness areas. As Abbott elaborates:

> Understanding the conservation estate as an idea (of wilderness) or as a processual landscape (of agency) greatly affects what can happen next. In the former the goal is to better conceptualise an "honorable, ethical, relationship with nature" – while in the latter the goal is to prompt a dialogue in which, through its articulation, wilderness and the conservation estate takes shape. In the former a sustainable relation with the conservation estate comes from right thinking while in the latter sustainability is an open-ended process derived through action.
>
> (2008: 155)

Abbott's approach highlights that landscape management can therefore be thought of as a series of experiments, and associated ongoing conversations, through which an appropriate landscape-specific approach is arrived at. Such an approach questions the primacy of a visual, expert-led version of landscape policy to make space for direct, lived environment experience. As Scott et al (2009) highlight, the dominance of visual-expert approaches to landscape management is being increasingly challenged. An attentiveness to these varied ways of knowing the world is now understood to be of increasing importance. As Nadia Bartolini and Caitlin DeSilvey remark, such a shift is increasingly necessary given the rapid environmental change impacting future landscape decision making processes. Thus,

> [e]mbracing opportunities that are afforded in the context of environmental change requires decision-makers, as well as the policy frameworks they work within, to be attentive to the varied ways that knowledge about landscapes is acquired, and to how different interest groups navigate change.
>
> (Bartolini and DeSilvey, 2021: 14)

Pause for thought: *Can you think of a landscape development you have encountered recently? How might an appreciation of how that landscape is "sensed" or known in "embodied" ways shape how it is managed?*

Collectively, these examples encompass numerous subjects and engagements, take in all manner of environments, and often deploy various ethnographic and participatory methodologies (see Chapter 13). They also make clear that, firstly, the cultural landscape is something perpetually in formation because it is constantly being made and remade; experienced, and re-experienced by different subjects (and bodies). Secondly, the cultural landscape emerges as the outcome of collective bodies (humans) engaging with each other (as well as the various kinds of nonhumans and forces). Whilst this opens a range of exciting opportunities for the landscape scholar, it is also important to consider the critiques of the phenomenological tradition (variously addressed or reproduced by contemporary work), to which we now turn.

Critiques, Response, and "Post-Phenomenology"

There are important areas of critique regarding the phenomenological approach to landscape. Like any conceptual framework, the approach foregrounds certain phenomena and questions about the world, whilst showing less interest in (if not excluding) others.

In terms of focus and attention, critics of practice-based phenomenological approaches argue that they can tend both towards the overly abstract *and* unhelpfully intimate. There are, for instance, works that pose major philosophical questions about the nature of experience, yet appear less obviously or directly relevant to managing landscapes or comprehending the everyday lives of inhabitants. Landscape research in this vein demonstrates, for critics, a level of "individual self-absorption" (Massey, 2006: 41) as it fails to engage with wider power dynamics, political contexts, or how landscapes are known in the realms of management or policy. Similarly, adherents to more critical approaches for cultural landscape study argue that phenomenological studies focusing on individual experience and sensory perception do so at the expense of discussing social, political, and cultural power relations (see Smith, 1979). Therefore, whether abstract or in attention to detailed lived experience, a loss of attention to structure, connections, and the conditioning of circumstance is seen to result.

An especially significant critique of phenomenological work concerns a sometimes-essentialist treatment of its (generally human) subject. Phenomenology (broadly conceived) "is often guilty of presenting an undifferentiated body subject" (Ash and Simpson, 2016: 60). That is, it assumes a person (the body-subject) simply exists in the way that they do, rather than as a product of history, context, or experience. The result can be a nascent tendency to universalise the subject, and to talk about "the body" in terms of *any or all bodies* rather than some particular body. Consequently, feeling that too little attention has been paid within phenomenological work to a diversity of bodies, their making and conditioning, has prompted efforts to extend its toolkit (see below). Such limited treatment of bodily difference comes in spite of much feminist theory and critique stressing the situated, specific, performative nature of identity and subjectivity (see Nash, 2000).

Finally, and relatedly, by working accounts through solely *human* experience – or a body that is entirely human – phenomenology risks "[reinstalling] human transcendence and [opening] up the old fault lines between humans and the rest" (Hinchliffe, 2003: 220). In this vein, another critique of the phenomenological subject has been a lack of historic attention towards how bodies take shape via relationships with other entities – be they other creatures or objects (like technology) – as "posthuman" subjects (see Chapter 10). Indeed, the sensing, phenomenological body that encounters landscape may not be (wholly) human at all.

Addressing Bodily Difference

Increasingly, landscape studies engage a range and diversity of bodies and bodily experience. Consider, for example, Hannah Macpherson's research with visually impaired hill walkers, which challenges the "latent ableism" (2009: 1042) of landscape studies by offering an alternative account of the Lake District as experienced by those afforded with different sensory capacities. Via interviews and ethnographic excursions with visually impaired participants, Macpherson explores how "body and landscape are complementary terms: each implies the other, alternately as figure and ground" (Macpherson, 2010: 4). For those who cannot (clearly) see their surroundings (or indeed those who can), an environment takes shape via processes of "intercorporeal landscape emergence" (MacPherson, 2009: 1044) as walking experiences occur through a host of relationships, "between guide and walker, through group practices in the landscape which stay with us in habitual ways, and through a historical relationship with one's own embodied past" (Macpherson, 2009: 1050).

In this way, MacPherson's work provides an early example of efforts to define a critical or post-phenomenology; seeking "to think about humanness without narrowing it to a holistic entity, to a transcendental identity that mirrors the root identity of the colonizing, Enlightenment man" (Norwood, 2018: 19). As Macpherson (2009: 1044) elaborates, traditional phenomenological accounts "prioritize the individual's embodied and habitual landscape experiences" without consideration for how they are shaped by social relations or wider cultural associations. Appreciating how a given subject's perception, relationship with, and experience of a cultural landscape reflect a situated, sometimes shared, context provides a route to placing that body within individual (or social) history ("the weight of numerous past associations" [Macpherson, 2009: 1052]) and the broader discourses, shaping how it is defined or regulated.

Thus, the body, involved with the world and the production of knowledge, is neither universal nor essential. A key aspect of critical or post-phenomenology, then, has been to locate embodiment within existing dynamics of marginalisation and inequality. To do so is crucial, as Eden Kincaid has argued, since subjectivity takes shape amidst sets of "dominant social meanings, cultural schema, and power relations" (2021: 307). These, in turn, affect how subjects relate or respond to *other* subjects. Various strands of "post-phenomenology" abound (see Kincaid, 2021; Ash and Simpson, 2016; Seamon, 2019), showcasing a diversity of interests and applications. Broadly, such work shares the sense that whilst "a good amount of phenomenological thinking is maintained", it is also "extended in critical, post-structural ways" (Seamon, 2019: 42). Ultimately, then, writing in this vein highlights that landscape can never be completely known because neither it nor the capacities of the body are static. Indeed, there are as many landscapes as there are bodies (see Box 8.3).

Box 8.3 Landscape's Embodied Potential

Embodied, practical conceptualisations of cultural landscapes consider encountered environments as "brought to hand" or "put to task". As geographer Mitch Rose argues, whilst landscapes appear as "definable material space", they are better understood as fields of potential: "the totality of possible performances imminent within [them]" (2002: 462–463). Thus, different *versions* of the world emerge through specific interactions, experiences, or uses.

This way of thinking about landscape is effectively illustrated via the writing of Scottish author and poet Nan (Anna) Shepherd (1893–1981). Her fiction, written during the early twentieth century, sees characters animated by encounters, sensations, and perceptions of scenes across her native Aberdeenshire. Indeed, her work speaks to much contemporary debate within the humanities concerned with human–environmental relations (Walton, 2020). Today, she is best known for her nonfiction account of the Cairngorm mountains of Scotland: *The Living Mountain* (1977). Popularised by modern nature writers, the book has become an exemplar of how literary attention to place can provoke ethical and existential questions of our situation amidst a living world (see MacFarlane, 2015; also Chapter 1).

In *The Living Mountain*, Shepherd renders the Cairngorms from multiple perspectives. Chapters variously entitled "Water", "Air and Light", "The Plants", or "The Senses" offer different "ways into" the mountains; revealing a cultural landscape remade anew by shifts in focus or bodily comportment (e.g. walking barefoot, paddling in ice-cold water, looking upside down). The Cairngorms as a field of potential is conveyed via her metaphor of "the total mountain". The sum of its multiple dimensions, it remains an ungraspable whole discernible only in slivers. As Shepherd suggests, her ability to comprehend this totality is thwarted by her bodily capacities ("If I had other senses, there are other things I should know"). Attempts to "know" the mountain are "endless": "the thing to know grows with the knowing" (Shepherd, 2011: 105–108). In other words, "[t]here are as many landscapes as there are combinations of bodies, matter and affective forces" (Garlick, 2023: 228). In Shepherd's writing, predating academic studies of cultural landscapes in (post-)phenomenological terms, we find engaging and readable accounts of a world "brought to hand"; remaking and escaping comprehension.

Chapter Summary

Box 8.4 Key Concepts

Chapter 8

The body
Phenomenology
Embodiment
Dwelling
Corporeality

This chapter outlines a phenomenological approach to landscape (Box 8.4). This conceptualisation of landscape contrasts sharply with much of what has come before in this textbook, in that it places bodies at the centre of knowledge creation. As Halling (2012: 1) observes, "phenomenology makes sense intuitively even though much of its language is esoteric": the emphasis in this work is upon the world as something always experienced and understood through the body. Landscape thus becomes a relational entity that comprises a particular meeting of sensing subject and physical world.

Whilst a number of important critiques concerning phenomenological approaches abound – a tendency to abstract away from lived experience, or conversely to become so enmeshed in intimate experience as to be granular in the extreme; and a sense in which one risks essentialising the subject apart from context or relationships – we encourage engagement with these ideas. Their applicatory potential is broad, and contemporary work has expanded and addressed some of these limitations, creating space within phenomenological thinking to accommodate a diversity of bodies and their situation within political and social power dynamics. Such work can prove valuable, showcasing theoretically and empirically how experience varies between people because of the bodies they have, but also how it is that specific experiences come to emerge where (and when) they do.

Suggested Seminar Activities

Activity 1: Read Chapter 11 ("The Senses") of Nan Shepherd's *The Living Mountain* (see Box 8.3). Although Shepherd, a teacher in Aberdeen, wrote *The Living Mountain* in 1944, it would take over 30 years for this book to be published. This short ode to a particular landscape in the Scottish Highlands is not self-proclaimed as a phenomenological writing, yet it can provide an entry into thinking phenomenologically about landscape and the ways in which "nature writing" attends to such themes.

Consider:

- How does Shepherd consider the body in terms of knowledge creation?
- Have you encountered similar writing about landscapes local to you?
- When was this book written (as opposed to published)? How does this fit with the timeline of thinkers encountered in this chapter?
- Read the following quote and discuss what Shepherd means by the final sentence in the Senses chapter.
- How might this book be re-written in respect to feminist and critical approaches to phenomenology?

Activity 2: Go for a short walk and pay attention to the way your body interacts with the environment around you. Keep a note of the experience. Use the following questions to prompt your noticing.

- How does it feel to walk on that particular surface?
- Does the surface change during the walk?
- What can you hear while walking?
- What can you feel other than the texture beneath your feet?
- Are there any particularly notable smells?
- Can you notice any impacts you are having on the world around you?
- How are the other bodies you encounter creating your experience?

Discuss these experiences in pairs or groups. How do they compare and what does this tell you about the landscape experience? The following scholarly snippets could be used to aid discussion:

- On encountering more challenging terrain movement inevitably slows, sometimes until "every step…become[s] a separate decision about direction and safety, and the simplest act of walking is transformed into a specialized skill" (Solnit, 2001: 133–134).
- "Belly laughs, titters, giggles, anxious exhalations, and guffaws would reverberate through the country air on days out walking, creating a transient sonic element of the landscapes we passed through" (MacPherson, 2008: 1081).

Recommended Reading

Ingold, T. 2011 [2000]. *The Perception of the Environment: Essays in Livelihood, Dwelling and Skill*. London: Routledge. Chapter 10. Also in paper form – Ingold, T. 1993. The temporality of the landscape. *World Archaeology* 25(2): 152–171.

Nettleton, S. 2015. Fell runners and walking walls: towards a sociology of living landscapes and aesthetic atmospheres as an alternative to a Lakeland picturesque. *British Journal of Sociology* 66: 759–778.

References

Abbott, M. 2008. Designing Wilderness as a Phenomenological Landscape: Design-Directed Research within the context of New Zealand's Conservation Estate. Doctoral thesis, Lincoln University.

Ash, J. and Simpson, P. 2016 Geography and post-phenomenology. *Progress in Human Geography* 40(1): 48–66.

Bartolini, N. and DeSilvey, C. 2021. Landscape futures: Decision-making in uncertain times, a literature review. *Landscape Research* 46(1): 8–24.

Cresswell, T. 2003 Landscape and the obliteration of practice. In: Anderson, K., Domosh, M., Pile, S., and Thrift, N. Eds. *Handbook of Cultural Geography*. London: Sage, pp. 269–282.

Daniels, S. and Cosgrove, D. 1988. Introduction: Iconography and landscape. In: Cosgrove, D. and Daniels, S. Eds. *The Iconography of Landscape*. Cambridge: Cambridge University Press, pp. 1–10.

De Certeau, M. 1984. *The Practice of Everyday Life*. Berkley: University of California Press.

Garlick, B. 2023. The total mountain: Nan Shepherd and the virtual qualities of landscape. In: Hall, M. and Hall, J. Eds. *The Mountain and the Politics of Representation.* Liverpool: LUP, pp. 211–231.

Halling, S. 2012. Teaching phenomenology through highlighting experience. *Indo-Pacific Journal of Phenomenology* 12: 1–6.

Hinchliffe, S. 2003. Inhabiting – Landscape and natures. In: Anderson, K., Domosh, M., Pile, S., and Thrift, N. Eds. *Handbook of Cultural Geography*. London: Sage, pp. 207–226.

Husserl, E. 1999. *The Essential Husserl: Basic Writings in Transcendental Phenomenology.* Ed. by Welton, D. Bloomington, IN: Indiana University Press.

Ingold, T. 2011 [2000]. *The Perception of the Environment: Essays in Livelihood, Dwelling and Skill*. London: Routledge.

Ingold, T. 2015. *The Life of Lines*. London: Routledge.

Jackson, J.B. 1984. *Discovering the Vernacular Landscape*. New Haven: Yale University Press.

Jackson, J.B. 1997. Landscape revisions. In: Horowitz, L. Ed. *Landscape Sight: Looking at America*. London: Yale University Press, pp. 333–371.

Jacobs, J. 1994. Earth honouring: Western desires and indigenous knowledges. In: Blunt, A. and Rose, G. Eds. *Writing Women and Space: Colonial and Postcolonial Geographies*. London: Guilford Press, pp. 169–196.

Kincaid, E. 2021. Is post-phenomenology a critical geography? Subjectivity and difference in post-phenomenological geographies. *Progress in Human Geography* 45: 298–316.

Lorimer, H. 2005. Cultural geography: The busyness of being 'more-than-representational'. *Progress in Human Geography* 29(1): 83–94.

Macfarlane, R. 2015. *Landmarks*. London: Hamish Hamilton.

MacPherson, H. 2008. "I don't know why they call it the Lake District they might as well call it the rock district!" The workings of humour and laughter in research with members of visually impaired walking groups. *Environment and Planning D: Society and Space* 26: 1080–1095.

MacPherson, H. 2009. The intercorporeal emergence of landscape: Negotiating sight, blindness, and ideas of landscape in the British countryside. *Environment and Planning A: Economy and Space* 41: 1042–1054.

MacPherson, H. 2010. Non-representational approaches to body – Landscape relations. *Geography Compass* 4: 1–13.

Mäkelä, H., Leiwo, L., Linkola, H., and Rinnie, J. 2024. The spiritual forest: An ethnographic exploration of Finnish forest yoga and the forest landscape. *Landscape Research* 49: 163–175.

Mason, O. 2021. A political geography of walking in Jordan: Movements and politics. *Political Geography* 88: 102392.

Massey, D. 2006. Landscape as a provocation: Reflections on moving mountains. *Journal of Material Culture* 11: 33–48.

Merleau-Ponty, M. 1962. *Phenomenology of Perception.* Trans. by Smith, C.. London: Routledge.

Nash, C. 1996. Reclaiming vision: Looking at landscape and the body. *Gender, Place and Culture* 3: 149–169.

Nash, C. 2000. Performativity in practice: Some recent work in cultural geography. *Progress in Human Geography* 24(4): 653–664.

Nettleton, S. 2015. Fell runners and walking walls: Towards a sociology of living landscapes and aesthetic atmospheres as an alternative to a Lakeland picturesque. *British Journal of Sociology* 66: 759–778.

Norwood, B. 2018. Disorienting phenomenology. *Log* 42: 11–22.

Relph, E. 2016 [1981]. *Rational Landscapes and Humanistic Geography.* New York: Routledge.

Rose, G. 1993. *Feminism and Geography: The Limits of Geographical Knowledge.* Cambridge: Polity Press.

Rose, M. 2002. Landscape and labyrinths. *Geoforum* 33(4): 455–467.

Scott, A., Carter, C., Brown, K., and White, V. 2009. 'Seeing is Not Everything': Exploring the landscape experiences of different publics. *Landscape Research* 34: 397–424.

Seamon, D. 2019. Wither phenomenology? Thirty years of environmental and architectural phenomenology. *Environmental and Architectural Phenomenology* 30(2): 37–48.

Smith, N. 1979. Geography, science, and post-positivist modes of explanation. *Progress in Human Geography* 3: 356–383.

Solnit, R. 2001. Wanderlust: A history of walking. London: Verso.

Thrift, N. 2008. *Non-Representational Theory*. Oxon: Routledge.

Thrift, N. and Dewsbury, J. 2000. Dead geographies – And how to make them live. *Environment and Planning D: Society and Space* 18: 411–432.

Walton, S. 2020. *The Living World: Nan Shepherd and Environmental Thought.* London: Bloomsbury.

Wylie, J. 2005. A single day's walking: Narrating self and landscape on the South West Coast path. *Transactions of the Institute of British Geographers* 30: 234–247.

Wylie, J. 2007. *Landscape*. London: Routledge.

Wylie, J. 2018. Landscape and phenomenology. In: Howard, P., Thompson, I., Waterton, E., and Atha, M. Eds. *The Routledge Companion to Landscape Studies*. London: Routledge, pp. 127–138.

9 Feeling the Landscape

Introduction: Dancing Lights

The Aurora Borealis, known as the Northern Lights, is typically observable in the Polar regions. Sometimes, though, it can fleetingly appear in Scottish skies. In the 1990s, Rachel recalls being hauled from her bed and traipsed to the fields behind the house to lie for hours with her family and watch this cosmic phenomenon. They were not alone; others from the village came with blankets and hot chocolate, or whisky. Those nights spent watching the "pale green and rose-pink [...] vast distant whispering swish" (Pullman, 1995: 183) felt special. The lights, the darkness, the lateness, the companionship combined to create an exciting moment. Even now, when the "aurora watch" app indicates a "red alert", this same atmosphere coheres with anticipation and memories of the past.

Such experiences have been commodified in recent times, as Gunnar Johanessen and Katrin Lund examine in relation to tourist encounters with the aurora in Norway. Guided experiences built around these light shows demonstrate the way in which a particular encounter with the landscape emerges out of "relational bundles" (Johanessen and Lund, 2017: 185) including tour guides, vehicles, other guests, weather patterns, solar particles, and darkness. As guides contextualise and build anticipation, a mood is choreographed and visitor experience stage-managed by tour companies. Yet such experiences are also emergent, contingent and, at times, elusive. Watching the Northern Lights is therefore not merely an act of seeing, but a way of *feeling*. The "lightscape" that tourists seek, like the atmosphere of the field behind Rachel's house, involves the lights themselves and many other elements besides to conjure "an environment of energies and capacities, a context within which a body feels and acts" (Edensor, 2010: 236). We might call this an emotional – or even affective – landscape, depending on our conceptual approach. It is to these terms and their implication for landscape studies that this chapter turns.

As discussed previously, the labels of non- or more-than-representational theory refer to various efforts since the late-1990s seeking to reconceptualise socio-cultural life beyond an erstwhile (over)emphasis on representations. That is, such work argues that cultural existence comprises more than the categories, labels, codes, or discourses that organise or regulate social relations and meaning-making. Crucially,

DOI: 10.4324/9781003172451-10

it includes "embodied gestures, actions and movements" as well as feeling, emotion, and sensation (Couper, 2015: 102). Indeed, scholars have argued that to focus only on representational practices and processes is to "deaden" our view of social life (see Thrift and Dewsbury, 2000) as various bodily practices, states, and intensities, never mind various lively *nonhuman* material processes and forms of agency (see Chapter 10), risk being left out. Non- or more-than-representational approaches, then, are an effort to reckon with "the necessary incompleteness (the lack) with the act of representation itself, that is, on that which eludes expression in language" (Smith et al, 2009: 12). This does not mean *ignoring* representations (e.g. visual imagery or language) which remain highly important for landscape (see Hunt, 2025), but rather appreciating how they incite or sit alongside myriad other dimensions of experience through which our lives unfold.

Where Chapter 8 focused on approaches applying phenomenology to landscape, centring the body and its capacities, here we consider the role of emotions and more nebulous felt-intensities, termed "affects", as they mediate and manifest encounters with or understandings of our surroundings. There are definite overlaps with previous discussion, particularly when considering emotion and affect in relation to bodies and their comportment. Nevertheless, there is also enough distinct about these concepts of "feeling" to merit a separate exploration.

Chapter Outline

This chapter is structured somewhat differently to others, spending substantial time establishing the terms "emotion" and "affect" upfront, before elaborating more their application to landscape. The first half of the chapter introduces and defines each concept, engaging the tensions, overlaps, and debates occupying the space between them. The latter half of the chapter then moves to spatialise and apply these concepts of feeling for the purposes of landscape study. Attention is paid to the notion of "emotional landscape" before turning to various spatial metaphors used to convey affective experience. In closing, we expand on the notion of "affective regions" as a way of combining previous insights, and attending to the diverse, historic, material, and emergent ways in which the qualities of landscape takes shape.

We acknowledge both that the ideas discussed here can be complex and confusing, and that many critics disavow such approaches as overly dense, unclear, or unhelpful. If you find these ideas challenging, you are in very good company! Regardless, we hope to demonstrate, document, and open up for scrutiny the many ways in which the world impresses itself upon us in visceral ways.

Theorising Feeling: Emotion and Affect

This chapter explores two ways in which cultural landscapes are "felt", *beyond* consideration of the body's physical capacities and sensory array (though both remain relevant). Firstly, via "emotion" and the ways in which emotional knowledge of the environment constitutes the cultural landscape. Secondly, through "affect", which connects a range of associated theoretical ideas associated with "non-representational" studies to expand the ways in which various felt "intensities" or states cohere into a landscape's "atmosphere", mood, or vibe. Elaborating each in turn, we reflect on the blurry and contested boundary between these concepts before proposing a pragmatic working-between as required by the specifics of (a) landscape.

Emotional Knowledge

Humanistic writing on place and environmental experience by the likes of geographers Anne Buttimer, Yi Fu Tuan, and Edward Relph, from the 1970s onwards, shows clear interest in human subjects' emotional knowledge of the environment. In part, an interest in such knowledge was explored as a means of expanding a sense of human–environment interactions beyond the heavily quantitative and abstract accounts dominating social science in the immediate post-war decades. Despite this early scholarship, much work on emotion in the social sciences at the turn of the twenty-first century could be described as marginal. "Emotional knowledge" was viewed as a vague or imprecise notion, irrelevant to research seeking to influence policy, politics or the "real world" (Anderson and Smith, 2001). That work which *was* done (as often remains true) tended to be under the auspices of feminist scholarship, exemplifying "a gender politics of research in which detachment, objectivity and rationality have been valued, and implicitly masculinized, while engagement, subjectivity, passion and desire have been devalued, and frequently feminized" (Anderson and Smith, 2001: 7). Indeed, such work often overtly resisted a conventionally held Cartesian dualism whereby "emotion" was positioned "'beneath' the faculties of thought and reason", thus tarred as unscientific or unscholarly (Ahmed, 2014: 3; also Jaggar, 1989: 152).

Nevertheless, as Anderson and Smith (2001) observe, emotions, emotional knowledge, and emotional labour are rarely purely subjective or private matters. Economic, political, and social relations are shot through with emotional content and mediated by changing emotional states. Moreover, as writers such as Ta-Nahesi Coates (2015) powerfully illustrate, in his case via reflections upon growing up Black in the face of racialised American policing and legal structures, the careful management of emotion often dictates how we negotiate (indeed, survive) different social contexts. Thus, feminist writer Alison Jaggar argues that emotions and emotional forms of knowledge sit centrally amidst socio-cultural relations as, "life without any emotion would be life without any meaning" (Jaggar, 1989: 161). All worldly activities are "steeped" in emotion, hence a need to consider their role in studies of encounters with landscape (Anderson and Smith, 2001).

Defining Emotion

To consider the relationship between landscape and emotion, this term requires definition. As feminist theorists have observed, however, doing so requires facing the "variety, complexity, and [...] inconsistency" in how emotion is characterised across everyday and scientific understanding (Jaggar, 1989: 153). Used to refer to a wide range of phenomena, emotion captures everything "from apparently instantaneous 'knee-jerk' responses [...] to lifelong dedication". It incorporates high-cultural aesthetic evaluations *and* "raw" experiential states (e.g. hunger, thirst, pain), both "background moods" – like "contentment or depression" – as well as situated "intense and focused involvement".

One way of distinguishing emotion from more general "feeling" or sensation is by virtue of its intention: it is directed towards, or responds to, specific phenomena or objects. Notable work by feminist philosophers such as Sara Ahmed and Lauren Berlant offers an articulation of our desires and engagements with the world in terms of the *promise* that certain objects (both real, tangible things *and* "ideas" or concepts) hold. In turn, certain things are culturally coded with the promise of "happiness" or fulfilling our desires (see Ahmed, 2010; Berlant, 2011). Thus, we feel "happy" towards, or in response to, some phenomena or encounter "judged" by the subject to be productive of happiness (or else sadness, fear, anger, etc.). In this way, emotions are differentiated from "non-intentional sensations" (Jaggar, 1989: 154 – see affect below).

Another way of defining emotion is as the socially constructed set of ways in which "feeling" is recognised, expressed, and managed as part of a cultural context and its conventions. Ahmed (2014) examines at length the *cultural politics* of emotion and the ways in which "normal" or "appropriate" modes of being emotional work through social assumptions, power relations, and hierarchies of feeling. We can also appreciate in, for example, the work of sociologist Arlie Hochschild (1983) how it is that emotions and expression (or suppression) in particular settings – such as the service industry encounters of commercial air travel – are part and parcel of the *work* that people do. To facilitate a particular customer's experience, an air steward undertakes emotional labour as part of their daily activities to promote and sustain a particular kind of encounter for the paying individual. As Jaggar observes, understanding emotions as socially constructed means appreciating them within geographical and historical context since "[n]ot only do some cultures identify emotions unrecognized in the West, but there is reason to believe that the concept of emotion itself is a historical invention" (1989: 153). She points to the "invention" of "Romantic love" in Medieval Europe – its associated icons and performances, and subsequent reworking across history and economic contexts – to demonstrate how the norms against which emotion is judged or incited (for example "norms about fidelity" and betrayal or the labelling of emotions towards certain situations or others as "outlaw" [Jaggar, 1989: 165]) are hardly settled or "natural". Far from "unmediated" experience, emotions are shaped by convention.

Between intentionality and social constructed-ness, "emotion" invokes the culturally situated and complex ways in which we and others express and comprehend our social contexts and relationships. Individuals (attempt to) control or manage their emotions in response to norms circulating through discourses pertaining to such feeling. Emotion thus names "ways in which we engage actively and even construct the world" (Jaggar, 1989: 159). They shape how we evaluate and perceive our world in a phenomenological sense: we *see* differently (through rose-tinted glasses, perhaps) when happy, sad, afraid, or otherwise emotionally charged. Ultimately, then, as Häyrynen (2021: 4) summarises, "emotions are key elements in the creation, interpretation and experience of space". Emotions matter in shaping our understanding of cultural landscapes as more than cultural categories for feeling, but also fully embodied states with implications for how perception and interaction with the environment unfold.

Theorising the Pre-Cognitive

Emotion constitutes an effort to make sense of or categorise "feeling" and our expression of it in relation to personal experience and cultural context. For philosopher Brian Massumi, emotion describes the "sociolinguistic fixing" of experienced intensity into a form. In naming a feeling ("I am experiencing happiness"), it is "owned and recognised" (Massumi, 2002: 28) and thus made *actionable* (e.g. "I will do this to make myself feel happy").

However, work within neuroscience and psychology has long problematised this view. The influential experiments of Benjamin Libet in the 1960s, 1970s, and 1980s suggest that, prior to one's mind manifesting the *intention* to act, the brain "ramps up" into a state of readiness preceding any conscious "action response" (see Fifel, 2018). For landscape scholars, this notion, "that consciousness is not instantaneous", poses "significant implications". As Hannah MacPherson (2010: 5) argues,

> "the half-second delay between stimulus and response means that we are not simply confined to a culturally constructed set of ingrained, reflective responses to landscape", and that" individual embodied experience [...] harbours the possibility of change`" and difference.

"Emotion" *can* imply conscious, cognitive processes of self-reflection and recognition that are incompatible with the idea that we are moved by the world in ways prior to conscious awareness. Consequently, since the late-1990s numerous theorists have deployed the concept "affect" to capture the immanent, pre-cognitive, "transpersonal" qualities of experience. In other words, affects name the surprising ways in which the world and its contents press upon, prompt, circulate, even "infect" different beings with feeling, sensation, or impulses to respond. Affect "displaces" the idea of emotion by "resolving [its] conceptual limitations [...] as an individual and personal experience or feeling" and avoiding (or refuting) the primacy of representation, language, and mental categorising when explaining our environmental interactions (Low, 2016: 152).

Defining "Affect"

It is important to appreciate that there are a great many ways of understanding, defining, and deploying affect. For influential theorist Nigel Thrift, it evokes "a set of concerns – with what bodies can do, with the power of emotions, with the crossover of 'biology' and 'culture'", and not a "finished analytic" (Thrift, 2008: 138). Nevertheless, we can, after anthropologist Setha Low (2016), identify two broad approaches to the concept. The first approach, evident in the work of Eve Sedgewick (2003), situates affect *within* the human subject's mind-body, drawing more substantially upon ideas from psychology and psychoanalysis to theorise it in terms complementary to a notion of "drives" (to eat, sleep, engage in sexual activity) as certain object-encounters stimulate a pre-conscious response. The second approach, more directly informing this chapter and work in landscape generally, locates affect *externally*, as a property of the world. It is this approach we discuss below.

An understanding of affect as a force (or set of forces) circulating through worldly relations draws from various philosophical precedents, notably Baruch Spinoza and the translation and application of his ideas by figures such as Gilles Deleuze. Affect, then, names "the push that keeps the world rolling over, the energy that fuels change" (Thrift, 2008: 113). As Seigworth and Gregg (2010: 1) elaborate, it constitutes "an impingement or extrusion" of relations as well as "resonances that circulate about, between, and sometimes stick to bodies and world", prompting or registering "the passage (and the duration of passage) of forces [...] that pass body to body (human, nonhuman, part-body and otherwise)".

Such language is complex, reflecting the challenge inherent in describing a concept intended to extend beyond the describable. In simpler terms, one might understand "affects" as all stimuli that prompt bodies (human or not) to viscerally feel *something* (whether un- or pre-consciously) and respond "prior to ideology" (Low, 2016: 151) and the codifying of feelings as emotions. As a non-cognitive "domain of unmediated experiential richness" and potential inherent to everything (Anderson, 2014: 59), affect exists prior to and separate from efforts to articulate or represent experience. When the world pushes on a body in some way, we might say that body has been "affected". The sensation of "being affected" constitutes "feeling", which, as it is evaluated, translated into representational terms, and categorised, becomes "emotion". Affect, thus, is the "motion of emotion" (Thien, 2005: 451). As geographer Derek McCormack (2008: 1827) summarises, "[a]ffect is a kind of vague yet intense atmosphere; feeling is that atmosphere felt in a body; and emotion is that felt intensity articulated as an emotion". In Box 9.1, we further articulate how affect can be grasped for the study of landscape via a focus on bodies' relationships with their surroundings.

Box 9.1 Affect and Bodies Out Hiking

Thinking about affect in relation to bodies can help to make it understandable. Affect helps us to consider what *bodies might be capable of*: how they are prompted, changed, or in some way moved by the environment.

The philosopher Gilles Deleuze, drawing upon Spinoza, argues bodies have the capacity to register the "push" of the world (its affects), as well as applying that "push" to affect other bodies. Thus, bodies have "the capacity for affecting and being affected" (Deleuze, 1988: 123). In this sense, Massumi (1987: xv) defines affect as something which triggers "the passage from one experiential state of the body to another" with consequences for what the body can do, or how it might act. So, whilst affects comprise a great many things – "properties, competencies, modalities, energies, attunements, arrangements and intensities of differing texture, temporality, velocity and spatiality" – they are *evidenced* by the way they "act on bodies, are produced through bodies and transmitted by bodies" (Lorimer, 2008: 552).

We could understand a landscape's "affects" by way of their *effects*, what landscapes do *to* the body. How do they press upon us? How do their sights, sounds, smells, feel, or vibe affect our state? And what ways do we press upon or change *them*? Affects can also manifest in or through *shared* experiences or relations. Furthermore, it is also possible for bodies to be more or less open to being "affected" by the world, more or less receptive of the landscape's push.

Consider a Fictional Group of Hikers Walking High in the Mountains ...

One group member is struggling, feeling exposed, cold, unfit, and unwell. Their bodily state changes as the environment and its materiality impresses (or pushes upon) the body in various ways: hard, steep ground; feelings of vertigo; bracing winds; altitude; sounds or glimpsed sights; the sense of disorientation or fear that comes with venturing into unfamiliar places. Equally, that struggling hiker presses upon – affects – their environment and other bodies within it. The wind refracts around their body; paths are eroded; their scent is carried on the air.

In addition to being produced by bodies, affects flow between them. They are transmitted, circulated, and contagious; potentially and frequently crossing between humans and nonhumans. One sullen hiker's struggle might even infect (affect) their peers, souring the mood as affect circulates, and changes through relationships (it is *relational*). But how this happens depends on how "attuned" (or available) these other bodies are. Another hiker in the party, wearing headphones and listening to music, is not aware of the struggles (and heavy breathing) of their companion. Neither do they have a fear of heights nor experience the cold in the same way.

Finally, we must appreciate that bodies, by way of habits, practice, or previous affective experience, will respond to a landscape differently. The capacity for landscapes to affect, and be affected, is historically contingent and often technologically mediated. Time spent acclimatising might result in a body less susceptible to altitude sickness, as can intergenerational inheritances for those who live in such places. Better-fitting shoes make one's feet less liable to blister in the presence of moisture, heat, pressure, and rough ground. Another member of the party, a seasoned mountaineer, is busy scanning the snow as training and prior experience tunes them in to the significance of how it looks, feels, and lies on the slope, inferring a risk of avalanche.

Overlaps, Tensions, Critiques

There is certainly much debate as to what, if anything, distinguishes emotion and affect, both ontologically (what they "are") and epistemologically (as forms of "knowing"). As suggested by discussion thus far, some scholars view emotion and affect as clear and distinct. Others, however, see a blurred line. Many (including the authors) feel the term one chooses when studying landscape is a pragmatic, context-dependent choice reflecting the language of research objectives, respondents, and relevant literature. Here, we touch on three areas of tension between these terms.

Emotion versus Affect

The nature of the relationship between affect and emotion is disputed. Approaches that build from psychoanalysis (e.g. Sedgwick, 2003) differ substantially from those framing affect as a circulating force. Thus, criticism of affect stems from limited engagement with key psychoanalytic concepts (like the unconscious), and apparent ignorance of how affect and emotion connect within studies of human psychology (see Pile, 2010). In response, those deploying a materialist, Spinozan notion of affect, view such critique as misunderstanding what the term aims to do. Intended as "a physical rather than psychological principle" concerning "movement or modification" affect untethered from human subjects (Dawney, 2011: 600) and concerns outcomes, or *effects*, more than thought. In this way, as Leila Dawney argues, emotion becomes *one way* (of many) of experiencing and expressing affective stimulus.

Practically, discussion of affective and emotional states can be hard to distinguish in research. For feminist scholar Liz Bondi, separating the two proves awkward in practice. Emotions too escape easy representation, are shared, and circulate. Writing about experiences of talking therapy practice for trauma, she notes, "because much communication happens outside of our conscious awareness, and because our affective boundaries are so permeable, there is ample scope for confusion about the relationship between our own bodily sensations and the feelings of others" (Bondi, 2014: 54). Therefore, whilst some propose clear distinctions, others see affect and emotion as similarly concerned with how bodies feel and register their situation, change states, and respond to the world. It remains crucial to define the terms being used, and how, in studies of landscape.

Hierarchies of Knowledge

Another debated concerns affect theory's tendency to ignore or diminish emotion as subordinate. Positioning affect "before" emotion reinforces a gendered politics of knowledge wherein masculinist, abstract affects, and forces drive a more grounded domain of emotion and experience. As Low describes, for example, emotion names an individualised "fixing" of free-flowing, creative "affect" force through categorising and representation (2016: 145). Thus, as "the labelling of sensation", emotion diminishes or simplifies the richness of experience extending "beyond" social convention (see Hemmings, 2012: 148). Interest in "affect" can thus imply interest in "more immediate embodied engagements [...] beyond language" (Smith et al, 2009: 12) before they become overlain by representational practices.

However, a "history of affect" that connects recent scholarship to longer lineages of feminist theory on feeling emphasises how concepts of affect and emotion have always enmeshed one another. Moreover, much feminist writing does not separate feeling and ideology, as work on affect can do (see below). Rather, ideological positions are understood to be steeped in affect and emotion: motored and mediated in myriad ways as feeling is, for example, "channelled within and across media with political consequences" (Tyler, 2008: 89). Ahmed's (2014) work, noted above, explicitly figures affect/emotion as part and parcel of "cultural politics", since who feels, when, and how reflects normative understandings, orientations, and performances of feeling that are power-laden and struggled-over. Whilst critical feminists may not reject affect *per se*, many resist its "post-ideological" formulations (Tyler, 2008: 90) and tendencies to hierarchise knowledge.

The Politics of Affect and Emotion

Debates abound over the critical and political implications of emotion and affect. Since emotion scholarship is open to questions of representation and discourse, it is sometimes framed as more politically relevant. Similarly, affect theory has been criticised for a politics that are opaque, unclear, or even absent (see Pile, 2010). Worse still, sociologist Imogen Tyler contends that some affect scholarship is guilty of a "celebratory" embrace of post-politics whereby accounts of cultural relations become technical matters of emergence, flow, and their workings-out rather than issues of power. Indeed, affect as an autonomous, pre-subjective force can seem to be "beyond power" and thus "complicit with [...] neoliberalism and individualism" (Tyler, 2008: 86). In turning away from methodological certainty or normative claim-making (i.e. distinguishing good and bad, right or wrong), whilst countering "the perceived limitations of ideological and representational critique" (Tyler, 2008: 87), affect scholarship risks over-theorising *what* happens at the expense of *why*, or in whose interests.

However, as many applying affect theory acknowledge, it *is* subject to manipulation and governance. Furthermore, the utility of affect is often asserted to lie in its enabling different kinds of analytical attention concerned with "what is at work", "what resonates through bodies as a result of their historical imbrications of material relations", and the character of those relations (Dawney, 2011: 599). Affect theory (and approaches characteristic of neo-vitalism more widely – see Chapter 10) represents a scholarly project interested in different questions, forms of critique, and conceptualisations of power (see Skiveren, 2023). In considering "the material production of bodies without having to posit any pre-social subject or body" (Dawney, 2011: 599), affect "decentres" accounts of individual subjects or objects "in" an environment or landscape and instead attends to how subjects and landscapes emerge *together*. The "political" questions it asks, then, are how do things happen, hold together, or change, and what are they capable of? The question of politics regarding affect and emotion is, ultimately, therefore a question of which aspects of landscape we seek to comprehend and change.

Spatialising Feeling

Emotion and affect are complex, debated, and widely applied terms. They can offer much to studies of landscape, and questions concerning how cultural relations with environments are mediated or motivated by feeling. Here we examine some ways in which emotion and affect can be articulated spatially, via an attention to emotional landscapes, spatial metaphors, and the notion of "affective regions". In this way, we connect these terms to scholarly interest in landscape and its more-than-representational dimensions.

Emotional Landscapes

In everyday speech, tacit links between environmental perception and emotional experience are often plain. Commonly, "we speak of the 'heights' of joy and the 'depths' of despair, significant others are comfortingly close or distressingly distant" (Davidson and Milligan, 2004: 523). As Davidson and Milligan assert:

> We live in worlds of pain or of pleasure; emotional environs that we sense can expand or contract in response to our experience of events [...] While we attribute emotional agency or capacity to a surprising range of external sources – saying, for example, that low clouds make us gloomy, while blue skies raise our spirits – our heart-felt articulations of emotion, themselves, acknowledge their interactional quality. [...] Without doubt, our emotions matter. They have tangible effects on our surroundings and [...] can clearly alter the way the world is for us, affecting our sense of time as well as space.
>
> (2004: 524)

Emotions can be made "articulate" through space. Yet, such talk is not mere metaphor. The way someone characterises, or acts in relation to, any environment is textured, or "complicated", by emotion (Low, 2016). Emotions and landscapes intermingle, shape, and influence each other and we can appreciate the "emotional geography" that results.

Emotional responses to places, landscapes, and environments shape our attachments to, or repulsion from, our surroundings. For example, an attachment to national or ethnic "homelands" involves understanding their *emotional* definition via "social and sensory relations" (Kearney, 2009: 211). Likewise, you might think of a landscape of childhood, or the various sites peppering the biographical journey to who you are now: a particular street or garden, a holiday location, the greenspace you sit amidst to escape the library at exam time.

Interest in emotions as part of the humanistic scholarship emergent the 1970s recognised, however, that such associations are both positive *and* negative. Tuan (1974) argues the significance of human–environmental attachments in terms of "topophilia", or place-love, and its antithesis ("topophobia", place-hate), as important in understanding the emotional lived experience of different environments. In later work, he (Tuan, 1979) elaborated how particular landscapes came to be constituted as "feared" or "fearful", harbouring either particular kinds of human or nonhuman material threats, uncertainty, or "chaos"; or by being culturally or mentally associated with phenomena that incited fear. Sometimes we retreat or flee from such landscapes; at others we seek them out as constituent of adventure or of difference. Feminist work has developed such insights with attention to the fear and anxiety shaping womens' experiences, such as a sense of safety on the nighttime route home through an urban landscape, to emphasise how such emotions reflect power and context (Pain, 2001). As discussed in Chapter 11, negative emotions associated with historic trauma, violence, and oppression adhere to urban and other landscapes, shaping their character in the present (Bordelon, 2022).

Emotional responses to landscape shape one's cognitive understanding within the encounter with it, as well as our anticipation of landscape, and how it is remembered. Being "slippery as objects of research, constructed, encoded, learned and mediated in culturally specific ways", yet also "spontaneous, uniquely felt and transculturally shared" (Häyrynen, 2021: 5), it is important to specify how and for whom emotional registers of landscape come to be known, shift, and change. Emotional landscapes are *temporal*: places once alive with the promise of happiness are later perhaps charged with loss, fear, or nostalgia (see Jones, 2015). Indeed, emotions of attachment or love, under situations of duress, may fuel anger or frustration if landscapes, or the ways of life they contain, are threatened. As Woods et al (2012) examine with regards British rural protests, deep emotional attachments to rural land and livelihood mobilised and justified a variety of protests between 1997 and 2002 against policy decisions by the then-UK government, including the imposition of hunting bans, controversial beef imports, and increased fuel duties. As the authors note, scenes of mass rallies in London, angry meetings of farming associations, and tractors blockading roadways across the country expressed both collective and personal, emotional attachments to the "British countryside" (manifest as real, lived environments and practices). Such emotions were both wielded by supportive organisations (such as the Countryside Alliance) and utilised in media coverage. Therefore, whilst "these emotions are not easily identified or expressed through language", they "exist in the iterative practices of rural life, in the senses of attachment and belonging to particular rural places and ways of life" (Woods et al, 2012: 573).

Spatial Metaphors for Affect

Like emotion, affect is a concept readily spatialised. This reflects a sense that affect is considered emergent within locatable contexts (such as landscapes) as the manner in which environments come to "feel", or the "mood" overlying them.

> ***Pause for thought:*** *Can you think of a time when you were in a particular urban landscape and some object or feature within the environment prompted you to feel, respond, or experience a sensation in some way? What do you think was the purpose of this from a design perspective?*

Several spatial metaphors have been applied to make sense of affect. Some articulate it in terms of "flow" or *pools*; as almost a liquid entity to be directed, channelled, or accumulated in reservoirs via effective landscape planning (for example, as in Tim Edensor's [2015] discussion of sports stadium architecture). This kind of liquid treatment of affect has also found purchase in landscape studies of illumination and darkness (see Morris, 2011; Edensor, 2012; Baker, 2015; Edensor and Lorimer, 2015), given the way that light "coats" surfaces, pools, or refracts around objects; mixing and layering landscapes to create a tone, vibe, or "atmosphere" by virtue of its presence or absence (see below). Edensor's work engaging the illuminated Blackpool (UK) seafront (Figure 9.1) recalls how colourful, seasonal light installations, together with sounds, smells, and lively onlookers, manifest "pools of affect within a seething space of movement and activity" (2012: 1111).

At the level of landscape, architectural interventions are often "designed to invoke affective response", as urban spaces are practically organised via "design, lighting, event management, logistics, music, performance, etc." to embed cues, images, signs, and technologies initiating an "affective response" in the subject (Thrift, 2008: 187). While the outcomes of this "landscape engineering" are "never guaranteed" (Thrift, 2008: 187), we might ask: how do planned landscapes prompt, create, or restrict the "flow" of affect; or even channel it towards certain locations, pooling a reservoir of sensation? Attending to contemporary examples of designed green landscapes in urban areas – the High Line (New York) and Under the Crystal (Copenhagen) – Casper Ebbensgaard (2017) discusses the creation of sustainable greenspaces on formerly derelict sites, and the creation of a particular sensory-affective space through curated encounters with weather, plants, materials, and other users. Thus, Ebbensgaard demonstrates how an atmosphere of urban calm can be created via landscape media that "instruct, choreograph and stage" a variety of human and nonhuman agencies, managing

Figure 9.1 Blackpool Seafront Illuminations in 2014, UK. Image by Lewis Clarke. Wikimedia Commons, CC BY-SA 2.0.

"performances and experiences" of "the sensation of nature" as it is manifested in city spaces (2017: 451).

Another spatial metaphor offered by anthropologist Kathleen Stewart (to whom we return below) is what she terms a "pocket". Characterising "a space opening out of the charged rhythms of the ordinary" (Stewart, 2011: 446), pockets articulate how an event or happening can almost create a temporary, bounded zone of intensity around itself. Elaborated as part of a fine-grained methodological attentiveness, elsewhere conveyed in her use of ficto-critical vignettes, the language of pockets offers a means of capturing the "ordinary" affective qualities of "what happens" (see Stewart, 2007) and the emergence, congealing, and dissipation of particular moments (see Stewart, 2012).

The work of Stewart has offered scholars of landscape tools for writing about landscape that employs noticing in reference to the small and continuous moments of affect and emotion. For example, David Crouch draws on her wider work to emphasise how "landscapes of leisure" are "unstable, contingent and fluid" (2015: 8), formed through the creativity of activities such as the sport or tourist recreation to exceed an account of landscape solely in terms of either "design" or the subject's interpretive "gaze". Landscapes and spaces of leisure emerge through lived encounters as "moments of practice" with different durations as remembered associations memory as "we lend character, meaning and significance to the materiality around us". As he writes:

> Pubs, dance clubs, the field of a car-boot sale, the beach, sports ground, heritage site, hills or gardens can be rendered in a particular character, influenced by our doing leisure. The spaces, their subjective, embodied feeling in us; the material world around us, objects and those living, combine in our feeling of space. What we find to be landscape is our expression of this mixture, an expressive poetics.
> (Crouch, 2015: 20)

By far and away, however, the most widespread spatial metaphor for the discussion of affect is that of *atmosphere.* Serving to elaborate its collective, immersive, circulatory, and "airy" properties, the notion of atmosphere, as geographer Ben Anderson (2009) suggests, extends the affect concept *beyond* a "push" upon individual bodies to consider how multiple affective intensities cohere into a situation or lived condition. Sometimes clarified as "affective atmosphere" (Bissell, 2010) – for affect and atmosphere are not synonymous terms – it might be thought of as the "moody force field" (Amin and Thrift, 2013: 161) enveloping or entraining bodies within a space to texture or produce how they feel. As Derek McCormack (2008: 413) explains, atmospheres are "distributed yet palpable, a quality of environmental immersion that registers in and through sensing bodies whilst also remaining diffuse, in the air, ethereal". In this way, they are *spatial* and potentially delineated by a sense of their edges, which remain always mobile, "amorphous and elusive" (Bondi and Davidson, 2011: 595).

Atmospheres emerge and circulate through a variety of stimuli and phenomena, including sound (see Box 9.2), bodily performances or gestures, representations and imagery objects, and other beings in an environment. As Edensor notes, we can readily imagine "the calm atmosphere produced through meditation, the animated atmosphere of a market or rock concert, or the sombre atmosphere of a gothic church" (2012: 1106). We can also, therefore, consider some of the individual sensations, phenomena, or affects that make up such atmospheres: music, air temperature, other sweaty or reserved bodied, lighting, time of day, anticipation. A politics of atmosphere also recognises their being cultivated, nurtured, and put to work (Thrift, 2008). As Martini and Buda write, "atmospheres, as elusive, affective networks, are neither silent nor neutral" (2020: 686). Thus, the many affects that *are* "a landscape", in the sense of stimuli "pushing" on any given body as they emerge from its various elements, come together in specific and contingent ways to differently constitute that landscape's "atmosphere", in the sense of it being a portion of the world with a particular feel or tone.

Importantly, whilst the above discussion defines atmosphere as "a space of affect" (Ingold, 2015: 73), some have argued such accounts figure it as "airless". Evoking more climatic associations of the term, recent years have witnessed a great deal of work conceptualising environments in terms of *volume* as airy, "elemental" domains (see Engelmann and McCormack, 2021). Consequently, Tim Ingold (2015) has argued that it is important to appreciate that the landscape is also a "weather world" – we inhabit, perceive with, and move through the "medium" of air, made known to us by virtue of weather and the other ways in which it presses upon subjects. Thus, "atmosphere" takes on both material (meteorological) and affective meanings (see Ingold, 2024), as well as being something that can both affect (e.g. manifest as rain, a cold wind, a clear, blue sky), be affected (winds diverted around bodies and buildings), and transmit affects (air-carrying sound, smell, or warmth).

Box 9.2 Affective Soundscapes

Discussion of affect in reference to cultural landscapes leads to consideration knowing environments differently, such as via sound. As Gallagher, Kanngeiser, and Prior, surveying the links between sonic sensibilities and affect, ask, "are the visual surface qualities of landscape, as perceived by a physically distant observer, all landscape is?" (2017: 623). An attention to sound, they reason, elaborates various ways in which sound and affect intermingle, producing a felt "soundscape".

The term "soundscape" was coined by R. Murray Schafer in the 1960s and developed by the World Soundscapes Project. Applied frequently in landscape studies, it articulates sound as central to our spatial perception. Indeed, landscape scholarship has explored the relationship between sound and perceptions of the environment in numerous ways (see also Chapter 5) to frame sound as a form of spatial knowledge and means of attending to more-than-visible phenomena (like the hum of an electricity cable as it crosses a landscape scene).

Of relevance to this chapter is the way in which sound might generate an affective atmosphere or press upon bodies, for "[s]ound, as physical vibration, is affective" (Gallagher et al, 2017: 625). As Gallagher et al outline, sound

> acts contagiously to modulate a dance floor, to repel bodies from alarms and sirens, or to innervate a wave of response during a vivid filmic scene. The affective aspect of sound comes precisely from the relations, exchanges and movements between bodies and environments.
>
> (2017: 625)

Gallagher et al argue that sound is important in understanding affective atmospheres because, firstly, it "moves through space" and with a "tendency to envelop other bodies". Secondly, sound *connects* bodies, as they collectively contribute as participants in the "sounding of worlds" (2017: 626), whether the sounds generated are perceptible by those involved or not, since we (and other creatures or materials) generate sounds all the time of which we are not aware. Sound therefore demonstrates that affect happens whether we cognitively involved or not. Sound has the capacity to impress upon bodies in pre-cognitive ways (e.g. low sounds producing queasiness) and thus sound constructs a felt landscape even when we "shut our earlids" (Schafer, 1977).

Politically, examining the ways in which sound is managed, encouraged, regulated, amplified, or directed can provide a way of apprehending how landscape's affects are choreographed to certain ends. A useful example that Gallagher et al discuss is one of the author's previous work on "high frequency devices designed to disperse young people" (Gallagher, 2016: 43) in public space, creating affective atmospheres thus exclusive of certain bodies. In turn, scholars of cultural landscape might enhance their "sonic sensibilities" (Gallagher et al, 2017: 618) by paying heed to the interlinking of "sound, space, infrastructures, matter and bodies" (2017: 625) as part of governance, regulation, and the excise of power.

Affective Regions

Having elaborated above some of the ways in which emotion and atmosphere are spatialised and known with reference to the manifestation of landscape, we close this chapter with discussion of the "regional" as an affective space, or landscape, with definable atmosphere – a "regionality" (Stewart, 2013). Elaborated via examples, we propose the affective region as a particularly useful, accessible, and holistic means of grasping the *feel* of the landscape.

Characterising affect at the level of the "region", or attending to how a sense of regionality or character is itself an affective atmosphere, can provide a way of accounting how it is that given landscapes emerged from, or define, particular "mood worlds" (Berberich et al, 2013: 319). A term proposed by Kathleen Stewart (2013), regionality develops her interest in American small-town life and its ordinary affective qualities as conveyed through everyday experience and daily practice (see Stewart, 2007). Thus, urban landscapes (like that in Figure 9.2) express a way of *feeling* and an emotional lifeworld, as much as a shared history, visual aesthetic, and agreed normative codes or conventions. The distinctiveness of a landscape is grasped through attention to its more-than-representational "regional qualities" (Stewart, 2013: 265), documented via vignettes of interaction and encounter that trace the "geography of what happens" (2013: 283).

The affective region or landscape, then, coheres from myriad "singularities" (Stewart, 2013: 283) that incite a sense of a region as having some kind of definable, shared feeling or coherence: sensations, objects, social relations, weather patterns, histories, and memories; entrained in laws, seasonal changes, colours, favourite beers, slang, local customs, prejudices, building styles, ways to prepare and eat specific foods. All come together as "a rickety infrastructure that makes perfect sense as a pieced-together, arcane affective atmosphere" and require a keen, ethnographic eye to document (Stewart, 2013: 280). Regionality is *felt and*

Figure 9.2 Walpole, MA – Kathleen Stewart's "ordinary" small-town landscape. Image by John Phelan. Wikimedia Commons, CC BY 3.0.

attuned to in particular moments: "the performance of the townie body" (Stewart, 2013: 282) via gestures, accents, or attitude; the visceral gut-wrench when one drives across the regional line. It constitutes a sense of landscape as what geographer Mitch Rose (2006: 538) terms (after Derrida) a "dream of presence": associations, experiences, and memories pulling together in a formation imbued with desires, fears, attachments, and senses of self and identity.

Box 9.3 Regionality in a Welsh Mining Landscape

Attention to landscapes as regional "mood worlds" reveals their temporal and material specificity: how history and the physical character of environments inform contemporary affective encounters or sensibilities concerning "identity", and their transformation over time. The work of Melanie Rhose, Rosie Day, and David Llewellyn (2020) shows how the atmospheric character of a former mining landscape and community in South Wales, defined by a form of energy economy and extractive industry, arises out of its material history.

They discuss the ways in which industrial activity bestowed characteristic affective material qualities to the landscape for residents of the town of Ynysybwl. During the height of the mining period, the air had "particular qualities imparted by dust and smoke" (2020: 243), with particulate matter carried across the land that would "literally colour their world" (2020: 243). Oral histories provided by residents recall a village with flowers that were "dark grey because of the coal dust flying everywhere" (Rhose et al, 2020: 143), and the thick, pervasive smell of coal smoke combined with a vernacular soundscape of industrial activity. More than this, residents recall "playful" encounters with the mines as children, inventing games and amuse themselves around these sites or on the spoil heaps as the landscape became "a playground". Reflecting on the range of responses and memories associated with the mine, the authors note the "diversity of atmospheres of varying temporalities in and around the mine, arising from the convergences and practices of different groupings of people in their encounters with the mine's materialities, and instigating outward affectual flows and precipitations of emotion" (Rhose et al, 2020: 142).

In contrast, today's partially recuperated, post-mining landscape is a place fundamentally different in its "feel". In the absence of mining, "key sensory cues ended, and people's experiences of their environment were transformed" (Rhose et al, 2020: 143). The authors note that the stories they have been told "tell us a lot about how change in energy systems feels" (Rhose et al, 2020: 144). The atmosphere that defined this landscape and region of Wales has changed dramatically, yet the traces, memories, norms, social relations, and emotions associated with that past remain part of it today. As Rhose et al note, then, reckoning with the centrality of extractive industry in this environment, and the influence it plays in shaping a sense of affective, regional identity, is fundamental when seeking to engage communities like that in Ynysybwl in processes of planning for, or implementing, new energy transitions and infrastructures. Indeed, "[s]uch endeavours may require [...us] to find ways of communicating affect and emotion, both in and about energy systems" (Rhose et al, 2020: 144).

Pause for thought: *Think of a famous or iconic landscape. What are some of the things that define its distinctive "feel"? Try to consider those phenomena that escape "representation" or are felt through their "press" on the body in some way.*

Regionality, therefore, has temporal stickiness (see Box 9.3). While the specifics of atmosphere shift, the feel of a region remains. In this sense, "regionality" gives us a means of attending to the lingering, persistent affective qualities of landscape, as much as their dynamism, effervescence, and emergence. An attention to affect, and how it works in landscape, then highlights the importance of the ordinary, the everyday and often overlooked in creating a sense of regional understanding over and through the passage of time.

Chapter Summary

Box 9.4 Key Concepts

Chapter 9

Feeling
Emotion
Affect
Atmosphere
Affective Region/Regionality

This chapter examined some ways by which we might attend to the "feeling" of landscape (Box 9.4). That is, the more-than-representational ways in which the environment informs, interacts with, or is changed via our emotions and affective state. Consequently, much of its discussion has been dense and theoretical. The conceptual framing of emotion and affect requires us to dig into ideas from fields as diverse as neuroscience, psychology, and continental philosophy. Equally, whilst "affect" and its associated terms can appear novel, it is important to contextualise the question of feeling within longer lineages of feminist scholarship arguing for an appreciation of embodied knowledges, including emotional knowledges, as well as less articulable feelings or states.

The account of landscape opened up by emotions and affect is one challenging some of the basic preconceptions of what constitutes a subject, agency, or how things can be considered to "happen". Affect, in particular, presents a view of the world alive with lines of force, intensity, and charge; pressing or pushing upon bodies (human and not) in ways that elicit some kind of response or cause them to change state and be made differently capable (see also Chapter 10). This is to invoke a world as process, made, and remade anew via events unfolding through relations, rather a sense of environments, societies, or subjects in terms of structures, representations, or self-aware agents acting in ways that reflect their regulation or adherence to norms. Affect suggests a more surprising, creative, unpredictable cultural landscape.

Ultimately, as is apparent from this chapter, "affect" is a term hard to describe or define abstractly because it aims to articulate that which is itself *un-representable!* As Wylie writes in his mid-2000s survey of landscape scholarship:

> Affect precisely describes domains of experience that are *more-than-subjective* and yet at the same time *formative* of senses of self – and formative, for example, of senses of self and landscape.
>
> (2007: 214)

So, affect comes before we even have a sense of ourselves as sensing being, able to articulate and make sense of the world … yet it also *drives* that very process of living, perceiving, becoming, and making sense. Whilst in Wylie's survey such engagements with landscape were largely "nascent", today they are anything but. Non-representational scholarship is a key area of development in landscape research over the last 20 years, engaging the world as a domain of sensation. As discussed above, the potential for landscape scholarship to spatialise and situate such experiences can prove incredibly valuable.

Suggested Seminar Activities

Activity 1: Go to a landscape and listen to the sounds that characterise it – as a class try to list all the sounds you hear in a 2-minute period to share in discussion (if you can't leave the classroom, then choose a landscape image and imagine yourself in that location). Then complete the following two, brief, exercises:

1. Play the following video (sound only) as loud as you can: www.youtube.com/watch?v=ghxzLw2wRis

   Close your eyes and imagine this to be the soundscape of the surroundings where you are. Consider:

   - *How does this new soundscape change the landscape? How does it make you feel? Does it provoke an emotional response that is different to the "actual" current soundscape?*

2. Put on headphones and listen to one of your favourite songs. You may pick whatever song you wish. Whilst you are listening to the song, pay attention to your surroundings. Consider:

   - *How does the experience of listening to this song, amidst this landscape, change it? Does it heighten or suppress any particular emotions?*

   In support of the above activity, you can read the following text in preparation or complement the activity with a discussion of the ideas therein:

   - Dodds, P. 2019. Hearing histories of Hammer Hill: Pop music as auditory geography. *Emotion, Space & Society* 30: 34–40. www.sciencedirect.com/science/article/pii/S1755458618301038

Activity 2: The philosophers Gilles Deleuze and Felix Guattari have suggested that an account of objects of existence can be offered by way of "counting" the things that affect it/us in some way.

Individually, or in groups, consider the "affects" (the "pushes" of the world that prompt a response) that characterise a landscape typical of your home "region". Try to compile a list, or perhaps a visual diagram, that records these affects, and then try to articulate something of the atmosphere, or "feel" of the landscape that results.

Recommended Reading

Edensor, T. 2015. Producing atmospheres at the match: Fan cultures, commercialisation and mood management in English football. *Emotion, Space and Society* 15: 82–89.

Stewart, K. 2013. Regionality. *Geographic Review* 103(2): 275–284.

Woods, M., Anderson, J., Guilbert, S., and Watkin, S. 2012. 'The country(side) is angry': Emotion and explanation in protest mobilization. *Social & Cultural Geography* 13(6): 567–585.

References

Ahmed, S. 2010. *The Promise of Happiness*. Durham, NC: Duke University Press.

Ahmed, S. 2014. *The Cultural Politics of Emotion*. 2nd Ed. Edinburgh: Edinburgh University Press.

Amin, A. and Thrift, N. 2013. *Seeing Like a City*. Cambridge: Polity.

Anderson, B. 2009. Affective atmospheres. *Emotion, Space and Society* 2(2): 77–81.

Anderson, B. 2014. *Encountering Affect: Capacities, Apparatuses, Conditions*. Farnham: Ashgate.

Anderson, K. and Smith, S. 2001. Editorial: Emotional geographies. *Transactions of the Institute of British Geographers* 26(1): 7–10.

Baker, J.C. 2015. Darkness, travel and landscape: India by fire-and starlight, c1820–c1860. *Environment and Planning D: Society and Space* 33(4): 749–765.

Berberich, C., Campbell, M., and Hudson, R. 2013. Affective landscapes: An introduction. *Cultural Politics* 9(3): 313–322.

Berlant, L. 2011. *Cruel Optimism*. Durham, NC: Duke University Press.

Bissell, D. 2010. Passenger mobilities: affective atmospheres and the sociality of public transport. *Environment and Planning D: Society and Space* 28(2): 270-289.

Bondi, L. 2014. Understanding feelings: Engaging with unconscious communication and embodied knowledge. *Emotion, Space & Society* 10: 44–54.

Bondi, L. and Davidson, J. 2011. Lost in translation. *Transactions of the Institute of British Geographers* 36(4): 595–598.

Bordelon, J. 2022. 'I dreamt about it, just like this, exactly the way it is': Haunting emotional geographies in Buenos Aires. *Emotion, Space and Society* 43: 100881.

Coates, T.-N. 2015. *Between the World and Me*. Melbourne: Text Publishing.

Couper, P. 2015. *A Student's Introduction to Geographical Thought*. London: Sage.

Crouch, D. 2015. Unravelling space and landscape in leisure's identities. In: Gammon, S. and Elkington, S. Eds. *Landscapes of Leisure: Space, place and identities*. London: Palgrave Macmillan UK, pp. 8–23.

Davidson, J. and Milligan, C. 2004. Embodying emotion sensing space: Introducing emotional geographies. *Social and Cultural Geography* 5(4): 523–532.

Dawney, L. 2011. The motor of being: a response to Steve Pile's 'Emotions and affect in recent human geography'. *Transactions of the Institute of British Geographers* 36: 599–602.

Deleuze, G. 1988. *Spinoza: Practical Philosophy*. Trans. Robert Hurley. San Francisco, CA: City Lights Books.

Ebbensgaard, C.L. 2017. 'I like the sound of falling water, it's calming': Engineering sensory experiences through landscape architecture. *Cultural Geographies* 24(3): 441–455.

Edensor, T. 2010. Aurora Landscapes: Affective Atmospheres of Light and Dark. In: Benediktsson, K. and Lund, K. Eds. *Conversations with Landscape*. Farnham: Ashgate, pp.227-240.

Edensor, T. 2012. Illuminated atmospheres: Anticipating and reproducing the flow of affective experience in blackpool. *Environment and Planning D: Society and Space* 30(6): 1103–1122.

Edensor, T. 2015. Producing atmospheres at the match: Fan cultures, commercialisation and mood management in English football. *Emotion, Space and Society* 15: 82–89.

Edensor, T. and Lorimer, H. 2015. "Landscapism" at the speed of light: Darkness and illumination in motion. *Geografiska Annaler B: Human Geography* 97(1): 1–16.

Engelmann, S. and McCormack, D. 2021. Elemental worlds: Specificities, exposures, alchemies. *Progress in Human Geography* 45(6): 1419–1439.

Fifel, K. 2018. Readiness potential and neuronal determinism: New insights on Libet experiment. *Journal of Neuroscience* 38(4): 784–786.

Gallagher, M. 2016. Sound as affect: Difference, power and spatiality. *Emotion, Space and Society* 20: 42–48.

Gallagher, M., Kanngieser, A., and Prior, J. 2017. Listening geographies: Landscape, affect and geotechnologies. *Progress in Human Geography* 41(5): 618–637.

Häyrynen, M. 2021. Emotional turn in the study of the environment and landscape. In: Häyrynen, M., Häkli, J., and Saarinen, J. Eds. *Landscapes of Affect and Emotion*.Leiden: Brill, pp. 3–19.

Hemmings, C. 2012. Affective solidarity: Feminist reflexivity and political transformation. *Feminist theory* 13(2): 147–161.

Hochschild, A.R. 1983. *The Managed Heart: Commercialization of Human Feeling*. Berkley: University of California.

Hunt, R. 2025. Bothy busi/yness: Navigating representation, practice and agency in the Scottish landscape. *Transactions of the Institute of British Geographers*. e70011.

Ingold, T. 2015. *Life of Lines*. Abingdon: Routledge.

Ingold, T. 2024. Landscape, atmosphere and the sky. *Lebenswelt Aesthetics and Philosophy of Experience* 23: 40–57.

Jaggar, A. 1989. Love and knowledge: Emotion in feminist epistemology. *Inquiry* 32(2): 151–176.

Jóhannesson, G. and Lund, K. 2017. Aurora borealis: Choreographies of darkness and light. *Annals of Tourism Research* 63: 183–190.

Jones, O. 2015. "Not promising a landfall ...": An autotopographical account of loss of place, memory and landscape. *Environmental Humanities* 6(1): 1–27.

Kearney, A. 2009. Homeland emotion: An emotional geography of heritage and Homeland. *International Journal of Heritage Studies* 15(2): 209–222.

Lorimer, H. 2008. Cultural geography: Non-representational conditions and concerns. *Progress in Human Geography* 32(4): 551–559.

Low, S. 2016. *Spatializing Culture: The Ethnography of Space and Place*. Abingdon: Routledge.

MacPherson, H. 2010. Non-representational approaches to body-landscape relations. *Geography Compass* 4: 1–13.

Martini, A. and Buda, D.M. 2020. Dark tourism and affect: Framing places of death and disaster. *Current Issues in Tourism* 23(6): 679–692.

Massumi, B. 1987. Notes on the translation and acknowledgements. In: Deleuze, G. and Guattari, F. 2013. *A Thousand Plateaus*. London: Bloomsbury, pp. xv–xviii.
Massumi, B. 2002. *Parables for the Virtual: Movement, Affect, Sensation*. Durham, NC: Duke University Press.
McCormack, D.P. 2008. Geographies for moving bodies: Thinking, dancing, spaces. *Geography Compass* 2(6): 1822–1836.
Morris, N.J. 2011. Night walking: Darkness and sensory perception in a night-time landscape installation. *Cultural Geographies* 18(3): 315–342.
Pain, R. 2001. Gender, race, age and fear in the city. *Urban Studies* 38(5–6): 899–913.
Pile, S. 2010. Emotions and affect in recent human geography. *Transactions of the Institute of British Geographers* 35(1): 5–20.
Pullman, P. 1995. *Northern Lights*. London: Scholastic.
Rhose, M., Day, R., and Llewellyn, D. 2020. Towards an emotional energy geography: Attending to emotions and affects in a former coal mining community in South Wales, UK. *Geoforum* 110: 136–146.
Rose, M. 2006. Gathering 'dreams of presence': A project for the cultural landscape. *Environment and planning D: Society and Space* 24(4): 537–554.
Schafer, R.M. 1977. *The Tuning of the World*. New York: Knopf.
Sedgwick, E.K. 2003. *Touching Feeling: Affect, Performativity, Performantivity*. Durham, NC: Duke University Press.
Seigworth, G. and Gregg, M. 2010. An inventory of shimmers. In: Seigworth, G. and Gregg, M. Eds. *The Affect Theory Reader*. Durham, NC: Duke University Press, pp. 1–25.
Skiveren, T. 2023. New materialism and the eco-Marxist challenge: Ontological shadowboxing in the environmental humanities. *Environmental Humanities* 15(2): 181–194.
Smith, M., Davidson, J., Cameron, L., and Bondi, L. 2009. Geography and emotions – Emerging constellations. In: Smith, M., Davidson, J., Cameron, L., and Bondi, L. Eds. *Emotion, Place and Culture*. Farnham: Ashgate, pp. 1–18.
Stewart, K. 2007. *Ordinary Affects*. Durham, NC: Duke University Press.
Stewart, K. 2011. Atmospheric attunements, *Environment and Planning D: Society and Space* 29: 445–453.
Stewart, K. 2012. Pockets. *Communication and Critical/Cultural Studies* 9(4): 365–368.
Stewart, K. 2013. Regionality. *Geographic Review* 103(2): 275–284.
Thien, D. 2005. After or beyond feeling? A consideration of affect and emotion in geography. *Area* 37(4): 450–454.
Thrift, N. 2008. *Non-Representational Theory: Space, Politics, Affect*. Abingdon: Routledge.
Thrift, N. and Dewsbury, J.D. 2000. Dead geographies—And how to make them live. *Environment and Planning D: Society and Space* 18(4): 411–432.
Tuan, Y.-F. 1974. *Topophilia: A Study of Environmental Perceptions, Attitudes, and Values*. New York: Colombia University Press.
Tuan, Y.-F. 1979. *Landscapes of Fear*. New York: Pantheon.
Tyler, I. 2008. Methodological fatigue and the politics of the affective turn. *Feminist Media Studies* 8(1): 85–90.
Woods, M., Anderson, J., Guilbert, S., and Watkin, S. 2012. 'The country(side) is angry': Emotion and explanation in protest mobilization. *Social & Cultural Geography* 13(6): 567–585.
Wylie, J. 2007. *Landscape*. Abingdon: Routledge.

10 Vital Landscapes

Introducing Landscape's Vitality

For conservationists and all manner of scientific researchers, the presence or absence of a major predator species, like the brown bear, has clear implications for landscape ecology. But how might bears being part of a landscape shape it in *cultural* terms? Finnish researchers Taru Pelota, Jari Heikkila, and Mia Vepsalainen offer an interesting account of how perceptions, and formations, of landscape in south-eastern Finland emerge out of the relationships between humans, brown bears, and waste disposal policy and infrastructure. Focusing on Moinsalmi, a village popular with second-homeowners attracted by the "wilderness" experience it affords, they discuss how such a *cultural* idealisation of this environment relies, partly, on the appearance of bears as icons of wild nature. Living in locations where humans are normally imagined not to be, the defining of the Moinsalmi landscape *as* wild depends on, or is "co-constituted" between, the assumed "placing" of bears and humans in different ways. The former reside in environments viewed as "undeveloped" or "more natural", which the latter seek out for tourism.

In 2007, human–bear relations became more fraught when the animals began exploiting the large communal waste containers used by residents as a food source. The failure of second-homeowners to follow restrictions around disposal of biowaste meant that, during the summer, "the table [was] set for bears" (Pelota et al, 2013: 467). Solving the issue proved complex because such waste units were part of a broader regional strategy to mitigate the impacts of increased tourism on wild places like Moinsalmi. This landscape was also subject to European protections relating to the conservation status of brown bears, with the region identified as a key breeding site and species stronghold area. Over the subsequent two years, the bears shifted from being exciting icons of wilderness to more often treated as pests. The animals, eating human waste, were held to intrude upon, and disrupt, the peaceful, predictable, pristine nature that visitors desired. By not adhering to these expectations, and being present in and around the waste containers, the bears were essentially "in the wrong place" (Pelota et al, 2013: 469).

This story demonstrates how *human* imaginations, understandings, and valuations of landscape are worked out in relationships with animals. Socially constructed representations of the environment (e.g. as "wild") hinge on nonhuman

DOI: 10.4324/9781003172451-11

presence (or absence). More than this, the Moinsalmi example highlights how *our* material modifications of the environment impact, or negotiate, other landscape inhabitants. Thus, attempting to understand landscapes like that of southeast Finland means examining relationships between humans, animals, and host of other, non- or "more-than-human" things (waste containers, fences, houses, even waste itself) besides. All these things, as Pelota et al conclude, are involved in "the material interactions which make cohabitation unsafe" (2013: 473). Efforts to understand, or intervene in, only one dimension in such complex webs of relations ignore both their complexity and the range of "actors" involved.

Across this textbook, we have considered landscape primarily as a manifestation of human cultural activities to understand how "we" transform, project meaning onto, or experience different environments. The (majority of) world that is outside of "us" has tended to sit in the background, featuring either as a *screen* (onto which our values, hopes, and ideas are projected) or a *stage* (upon and amidst which we act or adapt to fit the needs of our performances). Indeed, we have been at risk, after geographer Sarah Whatmore (2006: 603), of casting "the making of landscapes [...] as an exclusively human achievement in which the stuff of the world is so much putty in our hands". Often accounts of landscape presuppose separate, essential domains of "human/culture" and "nonhuman/nature", mixed together to make the world (see Hinchliffe, 2003). But such a way of thinking fails to acknowledge that a landscape, its composition, working, and histories, is more than just what we do to it, or what it means for *us*. Therefore, here we emphasise landscapes as an outcome of human and nonhuman relations.

Interest in this "vitality" of landscape builds upon work that conceptualises cultural and social relations as inherently more-than-human. In particular, the chapter draws upon "animal" (Philo and Wilbert, 2000) or more-than-human (Whatmore, 2006) geography research; theoretical and philosophical accounts of "vital materialism" and its version of a world constantly "becoming" (see Bennett, 2010; Deleuze and Guattari, 2013); and the wider "posthumanities", including work in history (O'Gorman, 2017), feminist theory (Haraway, 2008; Braidotti, 2014), and anthropology (see Kirksey and Helmreich, 2010; Hartigan Jr, 2021). Much such work echoes interest in landscape's "substantiveness" (see Chapter 3), as underpinning the critical Marxist and legal geography approaches of Ken Olwig, Don Mitchell, and others. Yet, it goes further, arguing that the land and its contents are not merely acted upon but *with*, being as the nonhuman world itself has "agency": the capacity to do things with implications for what happens. Despite being something long appreciated within non-Western ontologies of nature (see Chapter 7), the notion of nonhuman agency has become part of landscape studies significantly only since the latter 1990s, and we elaborate here the questions it raises.

Chapter Outline

This chapter considers how landscapes shape, and are shaped by, human relations with the nonhuman world. Importantly, this includes animals as well as a wider domain of "vital materiality". It considers how relations between humans and other things produce what some have characterised as the entangled, ongoing "event" of landscape (e.g. Massey, 2006; Garlick, 2019). The chapter begins by tracing early interest in animals in relation to landscape studies, before considering more recent relational and processual accounts of nature and materiality; including proposed means of theorising "agency" beyond the human. In closing, specific consideration is given to the study of landscape in the context of the "Anthropocene".

Animals and Landscape

There has long been appreciation for the role nonhumans play in the history of the landscape. Carl Sauer (see, e.g., Sauer, 1947, 1952), for example, noted the manner in which human culture, in transforming "natural" landscapes, affected the distributions of plants and animals, and their physical make-up (via domestication or breeding). Likewise, within disciplines such as geography there was an interest on the part of physical and environmentally minded researchers in understanding how distributions of animal (and plant) species had resulted from human activities, interventions, and changes to the environment as part of bio- and zoo-geography subdisciplines. Here, however, we elaborate the emergence of studies of animals in relation to the social construction and definition of landscapes; their meaning, governance, and "use".

Animals and Cultural Space

Amidst wider interest in questions of representation and environmental social construction (see Chapter 2), a number of humanities scholars turned to the "animal question" during the 1980s. Histories by the likes of Keith Thomas (1983) and Harriet Ritvo (1987) examined how, in Britain, human–animal relations evolved over time, tracking broader changes in environmental values, social norms, political philosophy, and scientific advancement. Likewise, geographer Yi-Fu Tuan's (1984) study of pet-keeping emphasised how certain spaces, like the home, came to be defined (and gendered) through emerging domestic human–animal relationships that associated pet-keeping with femininity. Collectively, such work made clear that human and animal interactions were both historically locatable and central to how different environments were perceived, organised, and engaged.

A strand of discussion within landscape studies through the 1990s took this interest forwards to consider animals within the construction and reproduction through images, texts, and discourse of identity, power relations, and imaginative geographies. In geography, in particular, several authors began to attend closely to the ways in which spatial orderings reflected and reinforced the category of "animal" in ways similar to the construction of other (marginalised) social groups. By "bringing the animals back in" (see Wolch and Emel, 1995) to accounts of space, place, and landscape, the myriad ways in which the environment was organised and governed by humans in different cultural contexts could be better illuminated. As political ecology put environmental issues into a socio-economic and political context, this new "animal geography" located understandings and interactions with animals within contingent cultural, political, and social world, as well as in relation to the definition and application of associated cultural categories (human, wild, domestic, feral, etc.) (see Anderson, 1997).

"Animal Landscapes"

Of interest to landscape scholars are the various ways in which landscapes are constructed, defined, and regulated as "animal spaces". That is, conceived, demarcated, or managed in relation to notions of where animals do or don't belong in the context of human values. Thus, the question of how we understand or organise the environment is entangled with animal presences – whether as natural inhabitants, or invasive pests. One might even appreciate how the atmosphere of a particular landscape is animated by animal presences in other ways; for example, vocalisations, smells, or embodied practices like horse riding, hunting, or birdwatching.

Consider also urban landscapes. In the nineteenth century across Europe and America, debates concerning animal welfare and sanitation reform saw cities increasingly "purified" as exclusively human spaces within public discourse. The bodily presence of animals driven to market, who might cavort and fornicate, or even be butchered, in the streets; and the smell and sight of offal and refuse from slaughterhouses discharged into gutters were seen to be detrimental to both health and morality in cities such as Chicago (Cronon, 1992). Simultaneously, other animals became *more* common in cities with, for example, the emergence of "the zoo" as a public institution to entertain, educate, and showcase creatures previously confined to the menageries of rich collectors (Anderson, 1995). Moreover, as pet-keeping became much more widespread, so domestic animals featured within urban landscapes, alongside codified conventions controlling how they did so. In the case of dogs, new spaces were created (dog parks or dog-free areas); practices and regulations introduced (the mandated use of leashes or muzzles in response to fears over bites or rabies); and new possibilities for conflict arose (the exasperation of those without dogs encountering dog waste) (Howell, 2015). Thus, one can appreciate the relationship between humans and creatures such as dogs as bound up with the historical emergence and formation of contemporary urban landscapes (see Pearson, 2021).

Beyond cities, scholars describe how nature reserves and wilderness refuges, national parks and landscaped estates, were likewise defined, visualised, and managed in relation to ideas of their proper or "natural" state that hinged upon certain animals being in or out of place. Often constructions of landscapes positioned them as *for* certain species (or not) Environmental management reflected and performed these assumptions, imaginaries, and knowledges of other creatures that circulated through numerous textual and representational sources such as literature, scientific articles, films, and artworks. As historian Emily O'Gorman (2017) describes with regard to her work on the early- to mid-twentieth-century management of water in response to fears over mosquito-borne malaria in Australia's Murrumbidgee Irrigation Area, New South Wales, such varied accounts of animal nature cohere into landscape's "imagined ecologies". These held understandings concern how nonhuman animals act, where or how they live, and the ways in which they might exert an influence upon those humans dwelling or transforming the environment.

There are numerous examples too of how "iconic" animals hold particular sway in the construction, management, and perception of landscapes in terms of their cultural, social, political, and economic significance. Ways of seeing landscape reinforce particular imagined geographies of human–animal relations, reproducing culturally and historically specific animal iconography (see Chapter 2). And yet, as practices of farming change, invasive biota spread, and experimental conservation initiatives propose the re-release of once-prolific species, so the conventional animal iconography of landscapes is also contested and transformed (see Buller, 2004). Sometimes there are active efforts in this direction: one might consider the ways in which conservation organisations – like *Rewilding Europe* – utilise images of restored landscapes replete with once-lost species like lynx and wolf as part of promotional materials, imagining the animals as *part of* such scenes prior to making this a reality via reintroduction programmes. In an example of ways of thinking about how landscape and animals are actively shaped for economic ends, Adam Keul (2013), considers the importance of alligators for swamp tourism in the Atchafalaya Basin, Louisiana. As he discusses, tour guides stage encounters with the swamp "wilderness" (see also Chapter 5) through acts of feeding, narrating, or advocating for the conservation of these reptiles that serve to draw visitors here. Thus, tourism is embroiled in acts of representation mediating perceptions of alligators and their environment amidst a wider context of a species' recovery from overhunting and domestic farming practices to serve the skin trade.

Animal landscapes, then, are spatial constructions concerning who (or what) belongs where, when, and under what circumstances. We accept the presence of (properly controlled) pet dogs in an urban square but would be understandably alarmed to encounter a lion, or a wolf, on the street. Equally, certain species once seen to be at home in the landscape become unwelcome as they are declared pests; or welcome once more, in the case of conservationist's efforts to encourage the reintroduction of lost species. Several studies of landscape have thus explored how such discourses of animal placing are encoded and practised within representations and acts of landscape(ing) (see Box 10.1; Proctor, 1998; Brown, 2015). As with other forms of "moral geography" (see Chapter 5), the reproduction of landscape is contested and variegated. The geography of animal presence, and associated practices or values (for example the permissibility of methods for killing or control), thus becomes a question of "spatial management" that works to "sidestep moral controversy" by literally zoning different human–animal relationships onto the land (Matless et al, 2005: 198).

Box 10.1 The Moral Geographies of Urban Avian Landscapes

The nests of the edible-nest swiftlet (*Aerodramus fuciphagus*) have long been prized as a culinary delicacy. Historically, its nests (Figure 10.1) have been harvested by hand, often by Indigenous communities in regions of Southeast Asia, such as Borneo, from the caves in which the birds nest. However, increasing demand and the economy for selling nests has resulted in the establishment of "swiftlet farms" in countries including Malaysia and Thailand over the past 30 years.

As Creighton Connolly describes, with regard to the UNESCO World Heritage City of George Town, on the Island of Penang, Malaysia, many swiftlet farms are found in urban areas as farmers repurpose old buildings as nesting sites. Examining this issue, notes Connolly, "renders visible the social structures over how the landscape is (or should) be made" (2016: 184) as different stakeholders marshal arguments for or against the industry, expressing perceptions of the birds as in or out of place.

Visible and prominent in the urban landscape, swiftlets do not remain within their farms. Moreover, "heritage" buildings appear to be more favourable nest sites and this is encouraged by farmers, for whom converting an old building is cheaper than constructing a new nest site or purchasing land outside the city. Newer buildings appear less attractive to the birds, perhaps owing to the different construction materials or modern protective treatments. Indeed, farmers have claimed that, since the birds are not "captive" as such, their presence in the city is "natural"; this place is their "home" and their existence is a historical phenomenon that predates their exploitation.

Figure 10.1 Edible Swiftlet bird nests. Image available in public domain.

Those against swiftlet farming cite public health concerns (such as the risk of transmitting viruses like bird flu) and the potential removal of George Town's UNESCO status. It is argued that the numbers of the birds found across the city exceed "natural" levels. Yet, even after state-led efforts to order the removal of the farms from inner-city areas, 2014 figures showed around 43 active sites in George Town, each accommodating around 1,000 birds (Connolly, 2016: 187). Whilst farmers assert the freedom of the birds to come and go, the use of recorded vocalisations to attract swiftlets to nest on specific buildings, and the proliferation of animals "[conditioned] to be reliant upon built structures for their survival" (2016: 188) is cited as evidence that this is very much a landscape outcome in which humans are culpable.

Thus, as Connolly describes, claims regarding the "naturalness" or otherwise of swiftlets viewed as in (or out of) place are framed as "a problem of scale" (2016: 188) as debate persists as to the right number of birds for this urban landscape. And, as noted above, the swiftlets *themselves* are "active agents" in this process: "the urban landscape is co-produced" through their interactions with farmers and other residents. The resulting landscape of George Town, then, "not only reflects the interests of humans, but also the actions of nonhumans" (Connolly, 2016: 189).

Transgressions and Agency

The question of cultures of landscape defined in relation to the presence or absence of certain nonhumans prompts us to consider that such entities do not always go – or stay – where our ideals or norms propose they should. Many nonhumans have the capacity to *transgress* "our" orderings: ignoring, being oblivious to, maybe even actively resisting or disobeying (thus raising the thorny matter of *intentionality*) human schemes. No matter our best efforts, fences fail, non-native species escape curated menagerie collections, and the material stuff of the world both assists and resists our imaginary boundaries and categorical designations of wild, domestic, or otherwise. For many, this raises the question of whether animals and other nonhumans might be understood as possessing *agency*: the capacity to act or, even, to be considered "subjects" (rather than mere "objects") in accounts of landscape.

A common proposition in contemporary studies of human–animal relations and posthumanist writing, the notion that animals have "agency" and might "resist" or act in ways driven by some form of intention, is not introduced without contention (see Philo, 1995). Whilst many Animal Rights advocates have long argued animals have interests and inner lives, therefore meriting ethical concern, those looking to attend to animal–human interactions have traditionally been less willing to decisively assert the existence of nonhuman agency. Such claims evidently risk the charge of anthropomorphism: the projecting of human traits or experiences onto other beings. To take the human as the yard-stick for interpretation in this way both risks ignoring fundamental *differences* between humans and other beings (or between organic and inorganic matter) and assumes (with some arrogance) that we can transpose our worldview onto others. Such a move can prove problematic, as appeals to "nature" to justify human activities operate on a similar logic, drawing equivalence to depoliticise and naturalise politically or socially contingent matters. Moreover, as feminists have argued, many scientists have been guilty of uncritically reading nonhuman relationships and interactions (such as those of primates) through the prism of human social norms (see Haraway, 1991).

To avoid the charge of anthropomorphism, some scholars opt to focus upon moments of "transgression". That is, the ways in which animals or other nonhumans act in ways that disrupt human-created boundaries or rules to create conflicts, prompt responses, or produce new kinds of landscape. This avoids having to ascribe intention (i.e. did the animal *mean* to "resist" this or that human activity), focusing on disruptive outcomes without adopting the full-blown language of "agency". Thus, the urban encroachments of the white ibis in Australia can be examined with an attention to how human practices and management strategies are changed by or respond to these birds (see McKiernan and Instone, 2016) without having to theorise those birds' "experience" of landscape or how it affects them. Nevertheless, as we explore below, there are those who find fertile (empirical and conceptual) grounds for considering the agents of landscape production as "more-than-human".

Landscape as "More than Human"

An interest in animals (and other things besides) as active participants in making worlds has increased since the turn of the century as many within the humanities and social sciences have engaged with conceptual work in science studies (such as Latour, 2005); feminist theory (such as Haraway, 2008); and strands of continental philosophy that variously emphasise a more complex and lively notion of the social and/or ecological. In such a context, geographer Sarah Whatmore argues the need for scholars "to reanimate the missing 'matter' of landscape" (2006: 603). Especially since her intervention, several landscape scholars, including cultural geographers, political ecologists, anthropologists, sociologists and historians, have showcased environments as hybrid, entangled, or otherwise "messy" in character. Whatmore (2002) herself offered an influential account of what she termed the "hybrid geographies" of humans and nonhumans as a means to propose an ontology (way of knowing) that focused upon relationships between different kinds of entities (themselves often the outcome of other relations) as the basis of existence. Thus, her account rejected the presumption inherent to many Western (and so-called modern) societies: that there exist "pure" domains of nature or culture (and their associated objects and forces), which, mixed together, create the world as encountered (see Chapter 1). Instead, she and others posit, everything is always already mixed up. The act of assigning things in the world to this or that portion of existence merely reflects cultural and political efforts at categorisation continually shown to be inadequate (see Latour, 1993).

As Whatmore (2006) observed, this "turn" towards material vitality that characterised scholarship in the wake of the millennium was better conceived of as a *return* to age-old questions: how does landscape get made? What does it mean? How does it work? Why does it change, or stay the same? For anthropologist Anna Tsing, examining human–nonhuman interactions requires taking "landscape as an object of analysis"; being as it names "the configuration of humans and nonhumans across a terrain" (2005: 173). Tsing's own landscapes of interest, characterised by intensive forestry, indigenous cultivation, mushroom picking, and extractive capitalism, are "weedy, patchwork" entities that underline how, despite what humans may do, they cannot control the other inhabitants of landscape, nor rid themselves of agents unaligned to our interests.

Relational Ontologies of Landscape

Contemporary work around the "more-than-human" often subscribes to a relational ontology. In other words, it begins from the premise that to understand existence (what "is") is ultimately a question of relationships: everything that exists or that happens is a consequence of things coming together, acting on or with each other, to produce some outcome. The role of the scholar is to fathom, trace, and reflect on the significance (in terms of historical transformation, causality, even implied ethics) of such connections, their manifestation, and where they might lead.

Several ways of theorising such an ontology have been outlined (see Box 10.2). Each approach emphasises different questions, or political or ethical concerns, and may derive from a distinct philosophical heritage or interpretation. What tends to *unite* these approaches, however, is appreciating that humans are not the only ones capable of acting or possessing agency. Rather, "how things happen" is understood as the consequence of complex interactions involving lots of beings, exhibiting different kinds of agency – or "acting" in different ways.

This expanded consideration of who acts is an important shift, because implicit or explicit discussion of "agency" in studies of landscape – as a capacity for (meaningful, or intentional) action shaping that landscape – has been historically anthropocentric (human centred). Whilst few would deny animals – not to mention rivers, rocks, plants, and weather systems – *do* things to (or in) the environment, their role within accounts of cultural landscapes has tended to be as things acted upon by us. From theorists such as Bruno Latour, Michel Callon, Donna Haraway, and John Law have come understandings of agency that more explicitly set out nonhumans as "acting" in important ways to enable, disable, or mediate (change) what happens as part of relationships, networks, or constellations of all sorts of different agents. Thus, we can appreciate how scallops influence the activities of fishermen and fishing policy (Callon, 1984); binoculars and bird books direct the doing of ornithological science (Law and Lynch, 1988); or speed bumps, sidewalks, and bollards create or delimit a range of behaviours within urban society (see Latour and Hermant, 2006).

Relational approaches which emphasise both humans and nonhumans as active participants in the world, then, promote consideration of landscapes as more-than-human outcomes of messy interactions. Their adoption within landscape studiescan be seen both in terms of defining what landscape "is" (as something made and defined through ongoing, amorphous relationships) and in thinking about how studies of landscape (and those engaging that term or its empirical referent in practical and policy contexts) contribute to its making (see Qviström, 2023). In Sarah Whatmore's terms, the entities, environments, spaces, or landscapes we encounter need to be understood as "a relational achievement": the outcome of numerous humans and, mostly (sometimes even exclusively – Lulka, 2009), nonhumans coming together to create new wholes (Whatmore, 1999). What's more, "humans" and "nonhumans" alike are, themselves, relational outcomes – a point central to much *post*human theorising (e.g. Haraway, 2008). A river system is what emerges from a combination of rocks, water, plants, fish, birds, soil, sand, weirs, boats, engine oil. A human body is the outcome of cells, metabolic reactions, bones, electrical signals, bacteria … As Whatmore summarises, a relational approach is a scholarly enterprise

> concerned with studying the living rather than abstract spaces of social life, configured by numerous interconnected agents – variously composed of biological, mechanical, and habitual properties and collective capacities – within which people are differently and plurally articulated.
>
> (1999: 26)

To know the world in terms of linkages and connections, then, rather than atomised, distinct beings moving across inert backdrops. It is a lively world defined by unique, impermanent, more-or-less consequential formations; a messy world that we might attempt to categorise (as "social" or "natural") which always exceeds or confounds such categories.

To summarise: relational ontologies are complex and much debated. A redefinition of agency as a capacity to enact change, mediate events, or more broadly make a difference (see Bennett, 2010) widens the scope of who, or what, is understood to matter within the accounts of landscape we offer. The nature of the relationship between entities becomes central to scholarly interest, with the world as both the context for, and outcome of, such relations (Allen, 2011). Whilst some work in this vein has been critiqued for *flattening* agency into something "mechanical", thus removing the "animality" from animal life (see Philo, 2005), there have also been examples of how work might situate and specify exactly what different actors are capable of in different contexts (Johnston, 2008; Barua, 2014). For example, attention to the ways in which animals sense and perceive their environments in ways that vastly differ from humans, and how such sensing is impacted by encounters with other beings, can invite speculation as to the more-than-human atmospheres of landscape (Lorimer et al, 2019). Such work represents an attempt to consider animals' own landscapes (or experiences of landscape), as we consider below.

Box 10.2 Relational Ontologies

Three Approaches

A range of "relational" approaches have been applied in landscape studies; offering different perspectives regarding how relationships form, endure, or affect each other, as well as conceptualisations of agency beyond humans. Below, we *briefly* introduce three influential approaches, with examples for how each has been applied to landscape study. There are other approaches besides these (Tim Ingold's [2015] notion of "meshwork" or Graham Harman's [2018] "object-oriented ontology" being two), so this overview is hardly exhaustive. The key point is that each approach has different underpinnings and offers some way of accounting for a landscape that is complex, messy, and continually being remade through both human and nonhuman forces.

Actor–network Theory (ANT): One of the earliest efforts to understand the world by foregrounding relationships, ANT emerged from sociological studies of scientific practice during the 1980s. Characterised as more methodology than conceptual approach, ANT figures the world as comprising networks of humans and nonhumans that stabilise into larger entities, themselves constituent of larger, further-reaching networks. Rejecting a notion of space as knowable in terms of measurable distance or volume, ANT advances a *topological* understanding wherein spaces (and landscapes) are "effects" of connections that are more or less intense, strong, or weak. Simply put, a landscape can be shaped, defined, or influenced by decisions or agencies located "far away" in terms of distance, yet which, for the purposes of producing outcomes that affect that environment, are more closely and directly involved than those physically proximate. As described by perhaps its most influential proponent, Bruno Latour (2005), the researcher's role becomes one of tracing a network of interest in relation to phenomena and understanding how it is produced, held together, or falls apart.

> ***Example:*** *Mattias Qvistrom's (2023) study of landscape architect's practices uses ANT as a means of following how "landscape" itself as an idea and material reality is constituted through the practices, objects, and relations of the field.*

Assemblage: Similar to some contemporary work in ANT, assemblage theory proposes reality as a heterogenous mixture of entities, themselves heterogenous in constitution. Derived from the philosophy of Gilles Deleuze (with Felix Guattari) (2013), assemblages evoke a more unstable, temporary, organic, and dynamic form of collective than a "network"; captured by the associated usage of terms such as "rhizome" or "refrain" to illustrate their formation and character. Endlessly connecting, interconnecting, forming, and breaking down, the entities which comprise assemblages are never fully actualised – their full range of capacities, or what they can "do", is never revealed – and, indeed, their "agency" may exert itself differently when enrolled into other assemblages. Thus, there is less certainty in assemblage theory as to how agency might be expressed.

Example: *Kaya Barry's (2019) study of tourists undertaking wilderness walks in Tasmania, Australia, explores the way in which an experience of a version of landscape emerges out of a constellation of objects, bodies, and the environment.*

Companion Species: An influential concept introduced by feminist theorist Donna Haraway (2008) to articulate a more-than-human ecology in terms of multi-generational, cross-species entanglement. Focusing on the case of humans and dogs, Haraway examines how these species have intermingled, historically, biologically, and culturally to propose that *all* beings "become" what they are through such "meetings" with others. Histories of any species or ecology are histories of "becoming with". Offering a situated, intimate, empirical, and ethically and politically alert accounting of relationships, Haraway's concept draws from wider feminist theory (e.g. Barad, 2007) emphasising what is termed "intra-activity": the fact that those involved in relationships are always also changed by them. For Haraway, the generative nature of relationships creates potential ethical obligations to others, as it also undermines any sense of humanity's boundedness or exceptionalism. Thus, the "companion species" idea is closely associated with "posthumanism": a collection of approaches that variously challenges notions of a stable, singular, superior human agent or subject as the basis for theorising knowledge, experience, or value (see Braidotti, 2014).

Example: *Anna Tsing's (2012) essay concerning the historical entanglement between mushrooms and humans emphasises how attending to "multispecies landscapes" as formed via relations of companionship provides a means of historicising contemporary environmental challenges and legacies.*

Beastly Places

There is today widespread scholarly recognition of what Henry Buller terms as "the independent materiality of animals" (2004: 132). Namely, we acknowledge animals' material existence apart from the meanings, associations, or categories we assign them. The expansion of agency via the theoretical frameworks outlined above provides a way of discussing landscapes as "beastly places": sites shaped and given significance by animals "performing their specific forms of agency [... and] creating their own worlds" (Philo and Wilbert, 2000: 20).

Thus, researchers deploy a critical, informed brand of anthropomorphism; one drawing from the work of biology and ecology to substantiate a potential sense of animal capacities, combined with the testimony, experiences, and ways of knowing with animals evidenced by those working with or alongside other creatures (Garlick, 2019). In terms of landscape study, one can appreciate how other creatures animate and perceive the environment on different, aligned, or overlapping terms to us. Equally, a sense that human management of the landscape places some limitations on, if not determining, how animals express themselves is also demonstrated (Clancy et al, 2021). Thus, scholars attempt to explain how "landscape" is textured by nonhuman experience and mediated by a host of other objects as forms of "multispecies collaboration" (see van Heijgen et al, 2024: 4) whereby different kinds of environments emerge as a result of different kinds of partnerships and involvements (see Box 10.3).

Box 10.3 Landscapes of Birds and Bird Hides

In 1957, a pair of ospreys (*Pandion haliaetus* – Figure 10.2) began attempting to nest on the shores of Loch Garten, in the Cairngorms region of Scotland. Wardens for the Royal Society for the Protection of Birds were keen to ensure these rare birds bred successfully and re-established the species after some

Figure 10.2 An osprey (*Pandion haliaetus*) alights on its nest in Aberdeenshire, UK. Image by Trevor Garlick, reproduced with creator's permission.

40 years of national extinction. They conducted the first of many round-the-clock monitoring efforts, guarding the nest against threats and observing bird behaviour to learn more about them. After ospreys successfully reared young to the northeast of Loch Garten in 1959, a wooden hide was erected annually, enabling subsequent wardens to watch the nesting raptors without disturbing them.

Reading the logbooks produced then, and since, within that and subsequent wooden hides, highlights how the landscape was organised around a problem of presence: "specifically, a tension arising around the articulation of human and avian presences to one other" (Garlick, 2019: 216). Humans needed to be close to ospreys, to protect and study them (with studies, it was hoped, informing better protection). However, initially, returning ospreys appeared skittish, alarmed at human proximity, and liable to abandon their eggs, or the nest site in entirety, if subject to excessive disturbance.

The bird hide provided a technological solution to this problem in the landscape. Yet, in its intended function, to "deceive" the ospreys into thinking that there were no humans present, it also serves to acknowledge these birds as sensing, perceiving subjects with agency in the landscape. As "a material intervention [...] infused with recognition of the animal gaze", the bird hide constitutes "a machine intent on reliably (re)producing particular kinds of animal–human landscape" (Garlick, 2019: 219). In addition, facilitating human "invisibility", the structure and contents of the hide at Loch Garten *prescribed* a particular kind of encounter with landscape. Containing logbooks, charts, and various visual aids for wardens to document the behaviour of ospreys, it reinforced ever more mechanical, standardised modes of perception.

However, understanding such a landscape only in terms of how humans perceived it evacuates the birds, and their perception, from the event. Indeed, the osprey logs document numerous incidences of noise emanating from (and beyond) the hide. The birds were clearly aware of humans coming and going. It seems that instances of human transgression into the area between the hide and nest, rather than human presence, elicited their distress. Therefore, reading these records against well-documented accounts of ospreys elsewhere acclimatising to human presence prompts an understanding of the landscape of bird conservation on Speyside as one of osprey *tolerance* rather than carefully managed obliviousness. Appreciating the role of ospreys as perceiving subjects reveals a landscape resulting from complex, cross-species negotiations, rather than merely humans acting "upon" the environment. In other words, "a more-than-human, phenomenal landscape in process" (Garlick, 2019: 229).

Pause for thought: *Can you think of an example of a landscape that is shaped by the activities of nonhumans in ways that evidences their ability to "make a difference" to that landscape?*

Vital Materialities

Accounting for landscape's "vitality" means acknowledging more than animals. In her book *Friction*, presenting an ethnography of logging across the sites and relations of globalised commercial forestry, anthropologist Anna Tsing observes that many accounts of "resource frontiers" render landscapes "inert [...] ready to be dismembered and packaged for export". In her account, by contrast, "the landscape [is] a lively actor"; "simultaneously natural and social" as it will "shift and turn in the interplay of human and nonhuman practices" (2005: 29). The construction of logging roads through Borneo's forests might create spaces of apparent "emptiness" (2005: 38) but none of these sites are ever, *actually* "empty". Weeds grow in and around the margins, and through the cracks of human infrastructure.

An interest in more-than-human dimensions of landscape proposes such engagements with materiality and substance (see also Chapter 3) with an appreciation that the "stuff" of landscape is, itself, lively and active. Political theorist Jane Bennett offers a particularly influential way of conceptualising the vitality of matter, organic and not, as "the capacity of things [...] not only to impede or block the will and designs of humans but also to act as quasi agents or forces with trajectories, propensities, or tendencies of their own" (2010: viii). Thus, all landscapes comprise innumerable "things" with properties and capacities to incite transformation in relations with other things, potentially across timescales that escape human understanding. Much work examining landscape in the way has developed over the last two decades with the application of relational ontologies (see above) to the theorising of matter and space. In Box 10.4 we consider an example from the writing of geographer Doreen Massey regarding the "event" of mountain landscapes.

Box 10.4 Thinking with Mountains

Mountains litter the imagination and have often provided a useful (conceptual) vantage for making sense of the world (Della Dora, 2017). Influential spatial theorist Doreen Massey, in her book *For Space* (2005) and a later essay (2006), offers a meditation on mountains for the conceptual perspective they afford, recalling a visit to the Lake District, UK. There, a landscape of glaciation, valleys and ridges, frames the surroundings. As Massey describes, one is tempted to view such a landscape as static and unchanging; particularly when looking upon something like Skiddaw, "a massive block of a mountain, over 3000 feet high, grey and stony; not pretty, but impressive; immovable, timeless" (2005: 131), looming over the town of Keswick.

And yet, Skiddaw, and its surroundings, reflects the "long and turbulent history" (Massey, 2005: 133) of geological activity: volcanic eruptions, glacial erosion, subsidence and uplift, weathering. They also evidence "a mobile geography" (134): the migration and interaction of continental plates. A night of pouring over geological information about the Lake District leads

Massey to reconsider this as a landscape of flux rather than an "achieved terminus" (137). Humans, animals, rocks, plants, and weather contribute to the "accretions of meetings" that make this scene. As Massey puts it:

> "here" is no more (and no less) than our encounter, and what is made of it. It is, irretrievable, here and now. It won't be the same "here" when it is no longer now.
>
> (2005: 139)

Thus, the Lake District showcases places and landscapes as "events"; naming a "thrown-together-ness" manifesting itself in "the unavoidable challenge of negotiating a here-and-now [...] which must take place within and between both human and nonhuman" (Massey, 2005: 140). Massey's writing on Skiddaw gives a concrete example of the vitality of landscape challenges to human sensibilities of time and space. It also reminds us that many of the "agents" making landscape are, after geographer Nigel Clark (2011), both "inhuman" and indifferent to our interests. What do the forces of geology, the nitrogen cycle, or river hydrology *care* of us? This doesn't mean that we are separated from such processes. Much of our social and political relations are profoundly *geological*, given the extent of our reliance on minerals and substances produced via geological processes. Indeed, the properties of such things, and how they act against or with us, are a key factor in making many landscapes what they are.

Anthropocene Landscapes

Any separate "human" or "natural" domains of existence appear difficult to maintain in our era of widespread ecological degradation, pollution, and climate change. Arguably, human influences upon the environment are such that indelible, anthropogenic marks are now inscribed into the planet's stratigraphic record. The argument that humans have even impacted *geology* fuels the assertion that we have entered a new geological epoch: the "Anthropocene" (see Lewis and Maslin, 2015). Whilst its definition and evidence base are much debated, possible starting points for the Anthropocene include the Columbian exchange (see Chapter 6), associated with European colonisation of the Americas; the industrial revolution and mass-burning of coal and other hydrocarbons; and post-war testing of nuclear devices in the Pacific. Moreover, some theorists suggest contemporary ecological transformations, whether associated with climate change, nuclear waste, or microplastic contamination, are so massive, distributed, complex, and non-linear as to render a "landscape perspective" untenable when trying to conceptualise human–environmental relations (see Morton, 2013).

However, there are also scholars who argue that, in fact, specific landscapes provide the most tangible frame for examining the Anthropocene, its characteristic phenomena, and the "everyday" ways in which it is experienced or comprehended (Fredriksen, 2021). Significantly, David Matless proposes the term "Anthroposcenic" to capture the ways in which certain landscapes (like eroding coastal regions in a time of rising sea levels and extreme weather) constitute the very grounds upon which our contemporary relationship to the environment is worked out, often in relation to deeper histories of climatic change. As he writes:

> The Anthropocene [...] has stuck precisely for its provocative conjunction of the human and the geological. In the same vein, the Anthroposcenic is a term through which landscape, in all its cultural complexity – material and imaginary, emotional and financial, immediate and intergenerational – might help visualise and represent a coming epoch.
>
> (Matless, 2018: n.p.)

One strand of Anthropocene landscape study has involved attention towards "blasted landscapes"; affected, or defined, by extensive levels of human-caused degradation, toxicity, pollution, or radioactivity. For Anna Tsing (2012), characterising the many ways in which (industrialised) society has left its marks on the world, such landscapes are neither devoid of life nor hope. In her own work on the networks and relations that characterise the matsutake mushroom industry (see Tsing, 2015), a delicacy that grows amidst the kind of disturbed ecologies left by intensive plantation forestry, she argues that considering the forms of life emerging out of such blasted surroundings reveals much about the more-than-human forces producing the shared, weedy, messy world to come.

Elsewhere, feminist science studies have argued that attention to toxicity can be productive for emphasising how human bodies exist in constant material exchange with their surroundings. As Stacey Alaimo (2010) notes, toxic bodies materialise the inseverable connection between our very corporeality and the environments we inhabit; as well as the capacities of various materials to infiltrate, attach to, and metabolise within our biology, creating something new. Studies of landscapes of toxic waste management, such as the Mina El Teniente copper mine in Chile, offer such insights and complicate any easy separation between the organic or non-organic; active or inert, as material properties change and reassert themselves in response to ongoing chemical reactions (see Ureta and Flores, 2018). Likewise, "nuclear landscapes" of the kind narrated by Rush-Cooper (2020) (see Activity 2) often reveal the complex material agency of dynamic, unpredictable, elusive matter at work in environmental encounters.

Summary

Box 10.5 Key Concepts

Chapter 10

More-than-human
Agency
Animal landscapes
Vital materialism
Relational ontology

This chapter has showcased ways of understanding landscapes as more-than-human (Box 10.5). Indeed, the empirical specificity of a particular landscape can mean that it provides an excellent lens through which to come to terms with the many human and nonhuman processes, relationships, and agencies that animate the world.

Drawing on approaches providing a relational understanding of the world, and foregrounding attention to the lively agency of nonhuman animals and other entities, this chapter continues the trend of the previous discussion (Chapters 8 and 9) by offering an account of landscape exceeding the representational. Approaches that emphasise textual interpretation, power dynamics, or discursive framings can fall short when it comes to the matter of how landscapes "are inhabited" in multiple, bodily, sensory, and affective ways; and (as above) when it comes to reckoning with "the inestimable vital qualities" of our surroundings (Edensor, 2022: 125–126). Whilst it is undoubtedly the case that human cultural acts of "placing" nonhumans are key in producing landscapes (as discussion of animal landscapes illustrates), it is also evident that nonhuman agency is fundamental to their formation and definition.

The chapter has noted the value of relational ontologies when it comes to thinking about landscape. Such conceptual tools foreground the heterogeneity of landscape as an ongoing, unsettled outcome of all sorts of different interactions. Indeed, encountering landscapes that help grasp the role of the often volatile, elusive, or persistent material agents that, in the course of industrialisation and scientific hubris, have been released into the environment suggests one route to better understanding and responding to today's environmental challenges (Matless, 2018). Taking seriously the kind of thinking elaborated in this chapter might lead to more positive forms of "landscaping" for the future. Practically, this might mean appreciating other creatures' affiliations with, or perceptions of, the environment so often excluded from conventional "Western" knowledge (see Chapter 7). It might also mean working within longer histories of landscape transformation to consider the (violent) impacts of present activities beyond short-term human timescales or priorities (see Nixon, 2011). Thus, attending to landscape relations reveals an unruly, complex, and lively world.

Seminar Activities

Activity 1: Look at the landscape visible from your window and try to make a list of all the human "actors" (planners, architects, pedestrians, etc.) who contribute to the scene you are seeing, hearing, smelling, feeling …

- *THEN*, make a list of all the other organic, animate beings (animals and plants) that inhabit the scene, and how they contribute to your experience of landscape.
- *THEN*, make a list of all the materials and objects that constitute the scene you can see. How does each one contribute to what you can see, hear, etc.?

Reflect on the above lists, how does expanding the appreciation of actors in the landscape have implications for:

- *The ethics of landscape* – what is right/wrong in terms of how we manage, interact with, or value this landscape; which aspects of the landscape have moral worth.
- *The politics of landscape* – whose interests matter in deciding the future of the landscape, how those interests are represented and communicated, as well as being weighed against each other.
- *How you understand, experience, or consider that landscape.*

Activity 2: Consider the following: Investigating nuclear landscapes is often a "tense" and difficult task, not least because such sites can be elusive; located out of sight, behind fences, or beneath a cloak of military secrecy (Pitkanen and Farish, 2018). And yet, because of the materials that are folded into the making of these landscapes, the stuff that comprises them cannot be considered settled or stable. Take the example of the Chernobyl Exclusion Zone: "an iconic landscape of toxicity, contamination and pollution" (Turnbull, 2020: 21) spanning an area of 4,200 km^2 in Belarus and Ukraine, including the entire (abandoned) town of Pripyat. The CEZ was established in 1986 following the explosion of one of the reactors at the Chernobyl powerplant. The reactor is today enclosed in an enormous protective concrete shell, given it will continue to decay for centuries to come.

Now read: Rush-Cooper, N. 2020. Nuclear landscape: Tourism, embodiment and exposure in the Chernobyl Zone. *Cultural Geographies* 27(2): 217–235.

Questions to consider:

- How is "radiation is woven into the landscape in material ways" (Rush-Cooper, 2020: 218)?

- In what ways does the Geiger counter constitute part of the experience of visiting the CEZ?
- Why might the nuclear landscape challenge the division of subject and world?
- How does the CEZ enable you to understand ideas of entanglement in the landscape?

Recommended Reading

Peltola, T., Heikkilä, J., and Vepsäläinen, M. 2013. Exploring landscape in-the-making: A case study on the constitutive role of animals in society–nature interactions. *Landscape Research* 38(4): 461–475.

Rush-Cooper, N. 2020. Nuclear landscape: Tourism, embodiment and exposure in the Chernobyl Zone. *Cultural Geographies* 27(2): 217–235.

Whatmore, S. 2006. Materialist returns: Practising a cultural geography in and for a more-than-human world. *Cultural Geographies* 13: 600–609.

References

Alaimo, S. 2010. *Bodily Natures: Science, Environment, and the Material Self.* Bloomington, IN: Indiana University Press.

Allen, J. 2011. Powerful assemblages? *Area* 43(2): 154–157.

Anderson, K. 1995. Culture and nature at the Adelaide Zoo: At the frontiers of human geography. *Transactions-Institute of British Geographers* 20(3): 275–294.

Anderson, K. 1997. A walk on the wild side: A critical geography of domestication. *Progress in Human Geography* 21(4): 463–485.

Barad, K. 2007. *Meeting the Universe Halfway: Quantum Physics and the Entanglement of Natter and Meaning.* Durham, NC: Duke University Press.

Barry, K. 2019. Material-bodily assemblages on a multi-day wilderness walk. In: Palmer, C. and Andrews, H. Eds. *Tourism and Embodiment.* London: Routledge, pp. 187–202.

Barua, M. 2014. Bio-geo-graphy: Landscape, dwelling, and the political ecology of human-elephant relations. *Environment and Planning D: Society and Space* 32(5): 915–934.

Bennett, J. 2010. *Vibrant Matter: A Political Ecology of Things.* Durham, NC: Duke University Press.

Braidotti, R. 2014. *The Posthuman.* Cambridge: Polity.

Brown, K.M. 2015. The role of landscape in regulating (ir)responsible conduct: Moral geographies of the 'proper control' of dogs. *Landscape Research* 40(1): 39–56.

Buller, H. 2004. Where the wild things are: The evolving iconography of rural fauna. *Journal of Rural Studies* 20(2): 131–141.

Callon, M. 1984. Some elements of a sociology of translation: Domestication of the scallops and the fishermen of St Brieuc Bay. *The Sociological Review* 32(s1): 196–233.

Clancy, C., Cooke, F., and Raw, Z. 2021. Entanglement, autonomy and the co-production of landscapes: relational geographies for free-roaming 'feral' donkeys (*Equus asinus*) in a rapidly changing world. *Geoforum* 123: 66–77.

Clark, N. 2011. *Inhuman Nature: Sociable Life on a Dynamic Planet.* London: Sage.

Connolly, C. 2016. 'A place for everything': Moral landscapes of 'swiftlet farming' in George Town, Malaysia. *Geoforum* 77: 182–191.

Cronon, W. 1992. *Nature's Metropolis: Chicago and the Great West.* London: Norton.

Deleuze, G. and Guattari, F. 2013. *A Thousand Plateaus.* London: Bloomsbury.

Della Dora, V. 2017. *Mountain.* London: Reaktion.

Edensor, T. 2022. *Landscape, Materiality and Heritage: An Object Biography*. Singapore: Palgrave Macmillan.

Fredriksen, A. 2021. Haunting, ruination and encounter in the ordinary Anthropocene: storying the return Florida's wild flamingos. *Cultural Geographies* 28(3): 531–545.

Garlick, B. 2019. Deceptive landscapes: Ornithological hide work and the perception of ospreys on Speyside, 1957–1987. *GeoHumanities* 5(1): 215–236.

Haraway, D. 1991. *Simians, Cyborgs and Women: The Reinvention of Nature*. New York: Routledge.

Haraway, D. 2008. *When Species Meet*. Durham, NC: Duke University Press.

Harman, G. 2018. *Object-Oriented Ontology: A New Theory of Everything*. London: Pelican.

Hartigan Jr, J. 2021. Knowing animals: Multispecies ethnography and the scope of anthropology. *American Anthropologist* 123(4): 846–860.

Hinchcliffe, S. 2003. "Inhabiting" landscapes and natures. In: Anderson, K., Domosh, M., Pile, S., and Thrift, N. Eds. *Handbook of Cultural Geography*. London: Sage, pp. 207–226.

Howell, P. 2015. *At Home and Astray: The Domestic Dog in Britain*. Charlottesville, VA: University of Virginia Press.

Ingold, T. 2015. *The Life of Lines*. Abingdon: Routledge.

Johnston, C. 2008. Beyond the clearing: Towards a dwelt animal geography. *Progress in Human Geography* 32(5): 633–649.

Keul, A. 2013. Performing the swamp, producing the wetland: Social spatialization in the Atchafalaya Basin. *Geoforum* 45: 315–324.

Kirksey, S.E. and Helmreich, S. 2010. The emergence of multispecies ethnography. *Cultural Anthropology* 25(4): 545–576.

Latour, B. 1993. *We Have Never Been Modern*. Trans. C. Porter. Cambridge, MA: Harvard University Press.

Latour, B. 2005. *Reassembling the Social: An Introduction to Actor-Network Theory*. Oxford: Oxford University Press.

Latour, B. and Hermant, E. 2006. *Paris: Invisible City*. Trans. L. Carey-Libbrecht. Available at: www.bruno-latour.fr/sites/default/files/downloads/viii_paris-city-gb.pdf

Law, J. and Lynch, M. 1988. Lists, field guides, and the descriptive organization of seeing: Birdwatching as an exemplary observational activity. *Human Studies* 11(2–3): 271–303.

Lewis, S.L. and Maslin, M.A. 2015. Defining the Anthropocene. *Nature* 519(7542): 171–180.

Lorimer, J., Hodgetts, T., and Barua, M. 2019. Animals' atmospheres. *Progress in Human Geography* 43(1): 26–45.

Lulka, D. 2009. The residual humanism of hybridity: Retaining a sense of the earth. *Transactions of the Institute of British Geographers* 34(3): 378–393.

Massey, D. 2005. *For Space*. London: Sage.

Massey, D. 2006. Landscape as a provocation: Reflections on moving mountains. *Journal of Material Culture* 11(1–2): 33–48.

Matless, D. 2018. The Anthroposcenic: Landscape in the Anthroposcene. *British Art Studies* 10. https://doi.org/10.17658/issn.2058-5462/issue-10/dmatless

Matless, D., Merchant, P., and Watkins, C. 2005. Animal landscapes: Otters and wildfowl in England 1945–1970. *Transactions of the Institute of British Geographers* 30(2): 191–205.

McKiernan, S. and Instone, L. 2016. From pest to partner: Rethinking the Australian White Ibis in the more-than-human city. *Cultural Geographies* 23(3): 475–494.

Morton, T. 2013. *Hyperobjects: Philosophy and Ecology after the End of the World*. Minneapolis, MN: University of Minnesota Press.

Nixon, R. 2011. *Slow Violence and the Environmentalism of the Poor*. Cambridge, MA: Harvard University Press.
O'Gorman, E. 2017. Imagined ecologies: A more-than-human history of Malaria in the Murrumbidgee Irrigation Area, New South Wales, Australia, 1919–45. *Environmental History* 22(3): 486–514.
Pearson, C. 2021. *Dogopolis: How Dogs and Humans Made Modern New York, London, and Paris*. Chicago, IL: University of Chicago Press.
Peltola, T., Heikkilä, J., and Vepsäläinen, M. 2013. Exploring landscape in-the-making: A case study on the constitutive role of animals in society–nature interactions. *Landscape Research* 38(4): 461–475.
Philo, C. 1995. Animals, geography, and the city: Notes on inclusions and exclusions. *Environment and planning D: Society and space* 13(6): 655-681.
Philo, C. 2005. Spacing lives and lively spaces: Partial remarks on Sarah Whatmore's hybrid geographies. *Antipode* 37(4): 824–833.
Philo, C. and Wilbert, C. Eds. 2000. *Animal Spaces, Beastly Places: New Geographies of Human-Animal Relations*. London: Routledge.
Pitkanen, L. and Farish, M. 2018. Nuclear landscapes. *Progress in Human Geography* 42(6): 862–880.
Proctor, J. 1998. The spotted owl and the contested moral landscape of the Pacific Northwest. In: Wolch, J. and Emel, J. Eds. *Animal Geographies: Place, Politics, and Identity in the Nature-Culture Borderlands*. London: Verso, pp. 191–217.
Qviström, M. 2023. Decentring landscape: Rethinking landscape analysis with a relational ontology. *Landscape Research* 48(8): 1108–1119.
Ritvo, H. 1987. *The Animal Estate: The English and Other Creatures in the Victorian Age*. New York: Harvard University Press.
Rush-Cooper, N. 2020. Nuclear landscape: Tourism, embodiment and exposure in the Chernobyl Zone. *Cultural Geographies* 27(2): 217–235.
Sauer, C.O. 1947. Early relations of man to plants. *Geographical Review* 37(1): 1–25.
Sauer, C.O. 1952. *Agricultural Origins and Dispersals*. New York: The American Geographical Society.
Thomas, K. 1983. *Man and the Natural World: Changing Attitudes in England 1500-1800*. London: Allen Lane.
Tsing, A. 2005. *Friction: An Ethnography of Global Connections*. Princeton, NJ and Oxford: Princeton University Press.
Tsing, A. 2012. Unruly edges: Mushrooms as companion species: For Donna Haraway. *Environmental humanities* 1(1): 141–154.
Tsing, A. 2015. *The Mushroom at the End of the World: On the Possibility of Life in Capitalist Ruins*. Princeton, NJ and Oxford: Princeton University Press.
Tuan, Y.-F. 1984. *Dominance and Affection: The Making of Pets*. New Haven, CT: Yale University Press.
Turnbull, J. 2020. Checkpoint dogs: Photovoicing canine companionship in the Chernobyl Exclusion Zone. *Anthropology Today* 36(6): 21–24.
Ureta, S. and Flores, P. 2018. Don't wake up the dragon! Monstrous geontologies in a mining waste impoundment. *Environment and Planning D: Society and Space* 36(6): 1063–1080.
van Heijgen, E., Driessen, C., and Turnhout, E. 2024. The landscape is a trap: Duck decoys as multispecies atmospheres of deception and betrayal. *Transactions of the Institute of British Geographers* 49: e12629. https://doi.org/10.1111/tran.12629

Whatmore, S. 1999. Hybrid geographies: Rethinking the *human* in human geography. In: Massey, D., Allen, J., and Sarre, P. Eds. *Human Geography Today*. Cambridge: Polity Press, pp. 24–39.

Whatmore, S. 2002. *Hybrid Geographies: Natures, cultures, spaces*. London: SAGE.

Whatmore, S. 2006. Materialist returns: Practising cultural geography in and for a more-than-human world. *Cultural Geographies* 13(4): 600–609.

Wolch, J. and Emel, J. 1995. Guest editorial: Bringing the animals back in. *Environment and Planning D: Society and Space* 13: 632–636.

11 Hauntings and Absence

Introduction – A Memorial Bench

Geographer Avril Maddrell recalls an experience waiting at York train station, UK; somewhere she has passed through many times. On this occasion, in the early 2010s, she noticed something: an "unusual bench" (Figure 11.1) on the platform. Its hand-carved oak design contrasts the metal seating around and, inspecting it further, Maddrell realises the bench is dedicated to a woman who died in 2005, aged 38. An inscription offers the name of the deceased and an epitaph ("still travelling"). As Maddrell observes, this bench reveals something to us about the involvement, past and present, of this woman in manifesting her current surroundings:

> No doubt the form, location and text of the memorial tell us something of the deceased's identity, but it also speaks of her vitality in the present, and her absence-presence in the lives of the bereaved.
>
> (2013: 502)

The memorial prompts some to reflect and feel something for this woman's absence. Perhaps her relatives visit the bench, sit here, and experience an emotional connection. Other people – busy commuters or tourists passing – probably sit here without registering its significance at all. Or, if they notice the inscription, perhaps they are largely unmoved by it. In short, this bench – as an object amidst a rather mundane urban landscape – moves people differently. Or, in light of Chapter 9, we might say it differently *affects* them. How much it does this depends upon how *tuned in* individuals are to its significance in relation to the absent woman who, it suggests, is still passing through this place on her continuing journeys beyond the land of the living. The affective and emotional quality of such an absence is differently registered, differently felt, by different people. Maybe, if the passing of this woman had been tied to some cause or other, such as a campaign to highlight the impacts of a particular illness or political issue, more attention would be paid and effort made to attract interest. Perhaps it is intended as personal and private, an unremarkable object to all apart from those who knew the deceased. However acknowledged, it changes – even if only slightly – one's experience of a particular urban landscape.

DOI: 10.4324/9781003172451-12

Figure 11.1 A memorial bench, York train station, York, UK. Photo by A.J. Smith.

This particular memorial prompts Maddrell to ask the following: how are experiences of landscapes in the present mediated by things not there – or no longer there? For geographer John Wylie, landscapes in his more recent writing are constituted, in part, by ghosts, hauntings, and other "ineradicable figures of absence" (2009: 283). For the purposes of this chapter, the bench raises questions about the *force* of memory, the past, or the absent – conceptualised by way of spectres, haunting, and ghosts – upon the present, and future, of landscapes. This in part reflects scholarly turns towards the "spectral" (influenced by the work of philosopher Jacques Derrida – see Luckhurst, 2002) over the past two or more decades within cultural landscape studies; as well as increasing interest in other forms of hauntings, voids, or absences as they shape spatial experience (e.g. Kingsbury and Secor, 2021). Here, we discuss how pasts and absences are variously folded into present landscapes, and how the resulting "hauntings" impact what we see, feel, or encounter.

Chapter Outline

This chapter applies some key ideas drawn from non-representational and affect theory (see Chapter 9) to expand on the notion of absence, and its importance for thinking about place and landscape. Specifically, it develops the concept of haunting, and the figure of "the ghost", to make sense of the ways in which cultural landscapes are experienced through absences, memories, uncanniness, and traumas. Examples of specific haunted landscapes, memorial practices, and ghostly encounters help illustrate the ways in which absence has implications for the politics and character of human–environmental encounters. In closing, the chapter considers how work exploring the "spectral" or haunted character of Anthropocene nature (see Chapter 10) provides a further area for cultural landscape studies to consider these ideas.

Uncanny Landscapes

The opening anecdote of the bench at York station evokes dimensions of spatial existence resisting easy articulation in language or other forms of representation. As explored across previous chapters, many facets of cultural existence can be understood as "more-than-representational": irreducible to categories, codes, or symbolic associations. These dimensions of experience concern bodies and their capacities, moods, feelings, sensations, emotions, and material agencies or forces. They are the sometimes less graspable, yet no less significant, traces of everyday lives and relationships producing the character of place (see Chapter 9).

Central to this example of the bench are the ways in which bodies register (are moved by) shifts in the emotional or affective "tone", or atmosphere, of surroundings; an atmosphere itself the product of all manner of charged relationships manifesting what we experience as a given, embodied "landscape" in a cultural context. There is also recognition that one needs to be "tuned in" to the bench. It is perceived as noticeable by those with some awareness of, or orientation towards, the significance or associations by which it is known and felt. Most usefully for this chapter, the bench prompts us to consider how applying ideas drawn from work on affect/atmosphere can help us grapple with how it is that things (or beings) which *are not* there continue to add to the "push" of the world (Thrift, 2008: 113) in a manner which might seem uncanny, weird, or eerie.

Uncanny, Weird, and Eerie Landscapes

The French aesthetic philosopher Jean-Luc Nancy, in an essay entitled "Uncanny Landscape", proposes that "a garden" be distinguished from "a landscape" by virtue of the capacity for one to "get lost" in the latter, but not the former. Putting aside the fact that gardens are often both landscapes and *landscaped* (see Chapter 2), Nancy's provocation usefully evokes something of the "limits" to our ability to comprehend a landscape, to grasp all of what it is, reflecting the way in which it exceeds that which we can represent in language. Such dimensions, perhaps, are that we encounter, or get lost in, when we meet with the "uncanny" in a landscape context.

The word "uncanny" is often used to talk about spooky, odd, or discomforting things. The dictionary definition indicates qualities of being "strange or mysterious, often in a way that is slightly frightening" (Cambridge Dictionary Online, 2024). When asked to contemplate an "uncanny landscape", one might think about experiences of (or in) an environment that disturbed, estranged, confounded, or unsettled. This is not least because much discussion of the "uncanny" within cultural theory is informed by the writing of the (in)famous psychoanalyst Sigmund

Freud. In his essay entitled "The Uncanny" (1919), Freud explores the concept – *unheimlich*, directly translated as "unhomely" – in relation to notions of repression and castration anxiety. For Freud, the uncanny is an experience associated with (unexpected) returns, repetitions, or doublings. It is not a case of direct encounters with the supernatural, but encounters *suggestive* of supernaturalism or which are difficult to fit within conventional schema. As the writer Mark Fisher summarises, for Freud, the uncanny denotes "the strange within the familiar, the strangely familiar, the familiarly strange" (2016: 10).

Examples offered in Freud's essay of things conjuring experiences of uncanniness include objects which seem human but are inhuman (such as prosthetic limbs, mannequins or waxworks, and automatons). Indeed, here we can reflect on Japanese robotics professor Masahiro Mori's (2012) concept of the "uncanny valley", originally proposed in 1970. Mori argued that we are particularly disturbed by objects which are so *nearly* lifelike in their appearance as to be akin to that which they simulate, yet remain just recognisable in their inhumanness that we know they are in fact simulations (e.g. CGI-generated animals or characters, humanoid robots). Based on such examples, literary theorist Nicholas Royle argues that the uncanny "involves feelings of uncertainty, in particular regarding the reality of who is and what is being experienced" constituting "a crisis of the proper [...] not simply an experience of strangeness or alienation" (2003: 1). Uncanniness might therefore accompany "something familiar unexpectedly arising in a strange or unfamiliar context", as it "concerns an estrangement of framing and borders" (2003: 2). Thus, an uncanny landscape might be a cultural landscape where things not there still exude an influence, where things appear out of place or awkwardly juxtaposed with their surroundings, or where questions of agency abound.

Seeking to further specify the character of uncanniness, Mark Fisher proposes two terms to both extend and nuance Freud's discussion: the weird and the eerie. Encounters with the weird are characterised as those with "that which does not belong" (2016: 10). Here, meeting some entity amidst the landscape immediately connects that landscape with other places, times, agents, or affects in a way that disrupts or disturbs the assumed boundaries of where we are. Fisher relays examples of objects or entities out of place, or of things colliding which should otherwise be kept apart conjuring that "sense of wrongness" (2016: 13). The eerie, meanwhile, usefully signals the uncanny way in which agency is expressed, specifically the question "who else is acting" (Latour, 2005: 43 – Chapter 10)? As Fisher elaborates, "the eerie is fundamentally tied up with questions of agency" and questions such as "[w]hat kind of agent is acting? Is there an agent at all?" (2016: 11). In this way, the uncanny, for Fisher, is heavily associated with haunting and a haunted world, where we are required to reckon with "the agency of the virtual" and "that which acts without physically existing" (2011: 18). In direct encounters with landscape, we are often encountering a field of potential – one version of an environment brought to hand out of many possibilities (see Chapter 8; also Garlick, 2023). In this way, examples of uncanny, weird, or eerie landscapes can emerge as different versions of a cultural landscape impinge on or muddle each (see Box 11.1).

Box 11.1 Weird and Eerie Landscapes

Weirdness on Ben Nevis

In summer, 2006, conservation volunteers working for the John Muir Trust (JMT) were carrying out a clean-up of Ben Nevis, the highest mountain in Britain. A popular tourist location, the heavy footfall of walkers (over 100,000 annually, according to the Trust) leads to erosion, litter, and efforts to maintain the landscape. In the process of their work, JMT volunteers dismantled a cairn (mound of stones) on the mountain's summit and made a startling discovery: an upright piano was buried beneath the rock (BBC News, 2006). Its origin was a mystery, though local residents in nearby Fort William claimed to know its provenance. Even having been the object of various challenges and publicity stunts over the centuries, the appearance of a piano on Ben Nevis was extremely *weird*. This thing was clearly out of place; its presence disrupted assumed boundaries between different cultural spaces (those for mountaineering versus the playing of piano music). It served to conjoin this landscape with a mythical history of strong-man contests and bizarre behaviours associated with the imaginary of "The Ben".

Eerie Nights in an Urban Park

Queens Park in Bolton, UK, was characterised by an unnerving and disturbing noise in late 2016. Residents complained that, in the hours of darkness after the park had been emptied and locked by council staff, a mysterious banging would echo from its interior. What was causing this noise? No-one could determine its source following initial reports in late November. One resident recorded the sound, sharing it with a local newspaper, and local authorities could not identify what caused it. Questions of agency abounded as to who or what was making the banging. The nature of the park landscape became subject to question until, eventually, after calls were made to Greater Manchester Police, it was determined the noise surely emanated from works being done on the nearby Bolton-Lostock railway. As a quirk of the landscape and its surface morphology, the sound of railway repairs was reflected, echoed, and made to appear as though it came from within the park (BBC News, 2016). The boundaries of Queens Park's sonic landscape were disrupted and appeared far more open-ended than one might have assumed.

Pause for thought: *Considering the examples above, have you ever had an experience of the weird or eerie? What was it, what prompted it, and how did it make you feel about your surroundings?*

For Fisher and Royle, the uncanny is productive as it questions the limitations of our (conceptual or material) borders (between space, time, forms of agency) and the ways we organise or portion up the world (e.g. the environment, history, or kinds of bodies). Uncanniness signals something happening outside of our normal schema, often attributed to the supernatural as that which does not fit within conventional understanding. For Fisher, the weird is "a signal that the concepts and frameworks which we have previously deployed are now obsolete", whilst the eerie can potentially "give us access to the forces which govern mundane reality but which are ordinarily obscured" (2016: 13). Together, these concepts suggest a need to attend to more than is apparent or straightforwardly *there* in the cultural landscape.

Nancy's (2005) suggestion that landscapes offer the possibility of "getting lost", retaining the capacity to unsettle or upset assumed demarcations, has been taken forward by other landscape scholars such as John Wylie (2016) to refute any sense of landscape as a stable site of dwelling or inhabiting. Rather, landscape is unstable, contingently held together, and embroiled with other spaces, times, affective states, and relations manifest as "hauntings" comprising "personal memories, collective histories, or physical remains" alongside which we abide (Paphitis, 2020: 342). Taking ghosts and haunting seriously amidst the cultural landscape study directs attention towards an important assumption: namely, that something must be *present* to affect.

Landscapes, Absence, and Haunting

It is only in recent decades that humanities scholars and social scientists have considered "ghosts" a relevant topic of study (see Luckhurst, 2002). In a short essay published in 2019, geographer John-Paul Jones III reflects on the value of ghost stories within his own field, cultural geography, and the challenge they pose for how we "know" the world. Retelling two ghost stories – one of a neighbour's encounter with a spectral child who once lived in their house; the other, relaying his personal experience of spirits passing through a friend's house during festivities for Day of the Dead (*Día de los Muertos*) – he emphasises the lived, material, and felt quality of hauntings. Such ghosts, he notes, often mark "their own boundaries of place"; appearing "place-bound" (2019: 281) in their activities. Certain sites, landmarks, and landscapes become meaningful via the work of ghosts remaking their boundaries and associations.

For Jones, attention to ghosts, spectres, and hauntings also has implications for epistemology (how we know, and what knowledge is) with regard to human–environmental interactions and their apparently intangible or seemingly absent dimensions. As he writes, responding to the question "[w]hat should we make of ghosts":

> The most comforting entry point is to recognize that, for one reason or another, many people believe in the haunting of everyday places. The retelling of these

> stories [...] can assuredly make our place stories more complex and textured, for they point us to those hidden corners of habitation that few people talk about.
> (Jones, 2019: 279)

The most mundane example of haunting might be the influence of *memory* on how we experience the cultural landscape, embodied in the memorial bench. A more provocative one might involve the ways in which heritage, tourism, or leisure industries actively trade on and evoke "haunted" and traumatic pasts to produce and commodify a certain kind of landscape experience (as in "dark tourism"). As sociologist Avery Gordon observes, "[h]aunting is a constituent element of modern social life" (2008: 7). For many people places, landscapes, sites *are* haunted. Crucially, such hauntings are both often "specific to place and landscape" and central to they get perceived (Paphitis, 2020: 343). Whether we believe in ghosts or not, we must account for this "haunting" in our studies of cultural landscapes.

Absences

Hauntings can be approached from the premise that experiences of environment are mediated by more than just what is tangibly present or *there* in the moment of an encounter. More specifically, events past or anticipated, and things now physically absent (or material traces of absence), often still have some kind of hold on us (Fisher, 2011). In terms of landscape, this is both a matter of materiality – how the "stuff" of landscape evokes past and future, absent and present – and a question of individual or social affiliations, memories, emotions, and affects. How might we reckon with the idea that the things that constitute landscape, and affect our sensibilities of it, include a whole range of things that have an influence by not being there? We need a way of comprehending the *presence of absence* as integral to the cultural landscape, and its potential, contested, and ever-changing meaning.

Indeed, for much of the textbook so far, we have been concerned with landscapes as a question of different kinds of presence. Landscapes appear as scenery upon which a viewing subject lays their gaze, the material setting for labours or struggles, spaces regulated through codes of conduct or enforced norms, or sites of bodily contact and sensory immersion. If we return to the bench opening this chapter, we can appreciate that its force in the landscape results from its capacity to confront us with something (someone) that is *not there.* This, as Maddrell has examined, is central to the emotional landscape of memorials and mourning wherein emotions of grief are "triggered" or worked out in relation to vernacular place-making practices that sustain and negotiate relationships with absent others (see Maddrell, 2009). In the case of the bench in the train station, we are invited to experience the urban landscape not through presence – things or beings in our surroundings that we perceive with and which, in various ways, serve to act on or mediate our experience – but through absence.

For several contemporary scholars, interest in questions of absence is allied with a broader project of theorising "the negative": the relations, forces, phenomena, or horizons of knowledge/experience/representation that resist easy accommodation within more "affirmative" accounts. In other words, those aspects of experience or processes that sit awkwardly within ways of thinking about human–environmental relations framing the world as a lively morass of potential and interaction. How, in one scholar's phrasing, do we "take seriously the imminent potential for cracking open, falling apart, slowing down, encountering limits, becoming numb or coming to a stand-still" (Dekeyser et al, 2022: 7) in our studies? How can we speak of "nothing" or nothingness in accounts that are always about "something", and how do we write about "the (im)possibility of relations between body and world" (Dekeyser et al, 2022: 16), accommodating such ruptures into the politics of landscape? These are complex and challenging questions that continue to be wrangled with in contemporary works.

In relation to cultural landscape studies, John Wylie offers a way into thinking with absence via an account of his visit to Mullion Cove, in Cornwall. Walking along the sloping ground atop the cliffs that encircle the harbour below, Wylie recognises how "the landscape we were looking from was already covered with eyes" (2009: 276) (see Box 11.2). Such absences can pull us out of the landscape – to other places and other times, dissipating and distributing the scene in front of us. Equally, absences might pull us in; creating a singularity or well in the landscape that we are drawn to explore, examine, and fill with stories, myths, or associations (Rose, 2010: 143). In this manner, attention to the absences of place constitutes a way in which landscape study attends to the spectral, ghostly, and hauntings that define or shape our world. The concepts of ghosts and haunting, in turn, focus our attention with regard to recognising landscape's *affective* qualities, and the ways in which other places, times, or agents bleed into our encounters with different environments; unsettling, transforming, or reworking them.

Box 11.2 The Benches at Mullion Cove

Along the cliffs at Mullion Cove, Cornwall, UK, John Wylie writes of finding a whole series of memorial benches, each affording a different view. Without initially realising it, he comes to appreciate the landscape as one made via a series of "looking-withs". He wasn't just seeing (and sensing) the cove via his own presence. Rather, he was seeing *with* (or through) the ghosts and memories clinging to the benches. He describes the palpable influence of those absent from the landscape, made present through the benches that conjured "a dense net of searchlights sweeping through the dark, sweeping over the waters" (Wylie, 2009: 277).

At the heart of each bench's point of view is an absence. Each bench is the material manifestation of love, memory, and death and conjures an affective connection with the past. This, Wylie argues, offers an important prompt to consider the experience of the cultural landscape because the majority of writing on the topic – including his own previous writing – emphasises landscapes as matters of *presence*, "defined above all in terms of contact, immersion and immediacy": a tangible, palpable, "matter of force, energy and process" that is very much alive (Wylie, 2009: 277). Yet the benches remind us of what is *not there*. It is not merely the case that these benches make those lost people present in some way. Rather, they direct our senses away from the landscape, to times, places, and beings not here. The memorialised person *infuses* the view from the bench as one is pushed to "remember via landscape" (2009: 281): "you nonetheless see them and see with them" (2009: 282).

Pause for thought: *Have you ever encountered a memorial bench, tree, monument, or structure in the landscape? How did this encounter change your experience of that place? Did it change how you perceived your surroundings?*

Hauntings and Ghosts

The question of how the past or the absent continues to act upon an experience of the present was one that concerned the later work of French philosopher Jacques Derrida (1994). Setting out his account of modern existence as haunted, he offered the neologism "hauntology" as a way of articulating this approach. Through an examination of the legacy of Communism, continuing to "haunt" Europe and its politics in the wake of Soviet collapse, Derrida alighted on the figure of the spectre or ghost, or *revenant*, which "begins by coming back" (1994: 11). Characterising this "haunted" character of modern life, the ghost is an entity signalling an experience of "time out of joint" (Derrida, 1994: 219). That is, encounters where past and present (or future: see Fisher, 2011), rather than remaining neatly organised within a linear conception of history, collide with and bleed into each other. The result is an unsettling, destabilising experience that draws us into (or out of) our situation and into alternative ways of being.

In his study of urban space informed by the psychoanalysis of Freud and others, geographer Steve Pile (2005) applies Derrida's thinking to conceptualise the city as a ghostly, haunted landscape. The urban realm, he argues, is filled with the ghosts of trauma, clinging to sites of past violence and our present-day associations with them (see Box 11.3). Thus, past intermingles with present, materially and affectively, throughout the modern city. Via different examples, including the defacement of war memorials and terrorist attacks in London, Pile argues for the influence of the past in producing what he describes as "a fractured emotional geography cut across by the shards of pain, loss, injustice and failure; an emotional world in which the ghost is the emblematic resident" (2005: 164).

In contrast to widespread cultural connotations with the ethereal or spiritual remnant of the dead (see Morton, 2015), Pile uses "ghost" to name a diverse set of material-affective forces mediating the perception of cultural landscapes and spaces. To encounter a ghost is, at its most basic, to feel the emotional or affective charge of the past or the absent through our bodies. Pile illustrates this through a discussion of war-time atrocities committed by the Japanese in Singapore, remembered via everyday practices of urban life and producing city landscapes thick with memories of violence and death and storied associations passed down by families through vernacular acts of memorialisation. Across the city he encounters places, feelings, and sensations that in often unconscious ways "alert us" to different sides of familiar landscapes. In another example, he discusses how the ghostly "presence" of fallen soldiers conjured at sites like the Cenotaph (a war memorial in Whitehall, Figure 11.2) in London makes such locations so potent. Consequently, acts which disrupt or dishonour their presence – e.g. acts of protest – provoke significant outcry. Thus, these "ghosts" (traces of different pasts), via their mobilisation in the present, are put to work asserting political values and circumscribing the dimensions of acceptable social action in the present.

Figure 11.2 The Cenotaph on remembrance Sunday in 2014, London, UK. Image by Harland Quarrington. Wikimedia Commons, OGL v1.0.

Box 11.3 Haunted by Fear in Buenos Aires

On an otherwise nondescript Buenos Aires street, a sign on one apartment bloc, Virrey Cevallos, advertises that it once housed a clandestine holding centre where political dissidents were tortured under the ruling dictatorship between 1976 and 1983. One of around 600 centres across Argentina, and 50 in the city, the sign situates this location within a wider network of places implicated in some 30,000 people "disappeared" as part of the "patchwork of terror" defining this period in state history. Often located in residential neighbourhoods, these centres manifested the regime's violent, oppressive presence in everyday urban landscapes. In turn, those living in these landscapes conducted themselves in fear and denial of what was happening behind the doors of the buildings they passed by daily.

Julia Bordelon has examined how the legacy of these centres in the present, and efforts to memorialise the atrocities associated with them, materialises and is understood in the context of a broader emotional geography of fear overlaying the cultural landscape of city (Buenos Aires) and nation (Argentina). Recuperated as "Spaces for Memory", such sites of trauma become centres for a transformative politics of *reckoning* which disrupts the persistent haunting of the contemporary city (and national memory) and the fixing of fear into their materiality. Drawing on Yi-Fu Tuan's concept "landscapes of fear", and elaboration via feminist debates concerning emotional knowledge (see Chapter 9), Bordelon examines how the recollection of historical threat, alongside the enduring traumas and violence tied to specific places in the city, imbued the landscape with a fear that had "a life on its own", outlasting the end of a dictatorship and subsequent (brief) efforts at reconciliation and criminal prosecution. As she remarks, "fear in this landscape had a remarkable endurance, representing a sort of purgatory that haunted the memory of Virrey Cevallos' neighbors" (2022: 6). Residual fear elaborates how traumatic pasts persist, expressing the "temporal fixity" of landscapes and a sense that "justice has not occurred" (2022: 4).

Whilst "neighbors understood the centre to be a physical site of terror", it continued to reproduce "a broader geography of fear" (2022: 4), within which the building remained a subject of suppression, denial, and limited political acknowledgement. The memory work undertaken by local activists at sites such as Virrey Cevallos (transformed into a Space for Memory in 2009) recovered local testimonies and resurfaced material traces of Virrey Cevallos' former usage to cultivate moments of emancipation "that can break haunting and fear" by bringing past traumas and histories to surface. Discussing entries in the centre's visitor log, the work of those managing the site to recover material traces of its history, and the way in which memorial work has re-presented the building to the community, Bordelon notes how the site has now become a place for confronting how a pernicious culture of silence ensures that the landscape *remains* haunted. As she argues, this work allows "[t]he reintroduction of time" to the city's landscape alongside "emotions other than fear" (Bordelon, 2022: 6), assuaging the "disorientation" experienced in relation to unknown (or unacknowledged) atrocities and enabling fear in the neighbourhood to be "laid to rest" (2022: 6).

Haunted Landscapes

For geographer Karen Till, writing on the memorial spaces that characterise the heart of a reconstructed, post-war Berlin emphasises that *every place* is more or less haunted:

> places are never merely backdrops for action or containers for the past. They are fluid mosaics and moments of memory, matter, metaphor, scene and experience that create and mediate social spaces and temporalities. Through place-making, people mark social spaces as haunted sites where they can return, make contact with their loss, contain unwanted presences, or confront past injustices.
>
> (Till, 2005: 8)

Each landscape is textured by many affects. Memory, and a sense of belonging, articulated with reference to the landscapes we have encountered or made our home over the course of our lives, inform subsequent encounters, here and now (see Tolia-Kelly, 2016). Other absences or associations can surprise or force themselves on us in the form of ghosts that see the past and present bleed into one another. In other instances, the present is haunted by the hoped-for promise of a different kind of future, perhaps one that failed to materialise or is continually deferred to justify political activities in present landscapes (see Kemmer and Simone, 2021). As Owain Jones writes, in a reflective essay concerning his personal links with the landscapes of Britain's Severn estuary:

> The past survives however much one tires to drive it down and away from one's consciousness. It rears up provoked by something overheard or a scene, a place, an object, a tune, a scent even.
>
> (Jones, 2015: 1)

Pile proposes that certain ghosts demand "a particular kind of seeing" (2005: 139). In the case of cultural landscapes, one might know *this* was the place where something happened. At other times, perception is actively directed towards a site of tragedy, trauma, or remembered significance. Understood this way, a practised and practical, embodied attention or orientation towards ghosts makes clear a more richly patterned emotional geography of the world:

> Ghosts disrupt notions of linear times and spaces: by returning, they alter both time and space, leaving neither quite where they had seemed to be. Ghosts, we can say, haunt the places where cities are out of joint – out of joint in terms of both time and space. They join the living world through fissures, sometimes these are called "anachronisms", sometimes "ruins": cities are full of them. They grip the imagination, screaming at the trauma and pain they have had to endure, lamenting the injustices that have befallen them.
>
> (Pile, 2005: 164)

Maddrell demonstrates this activation of landscape haunting via studies of bereavement and memorialisation. She describes how memorials, including the bench in this chapter's opening, produce landscapes where things absent are a constituent part of the character of place. These absences are rendered tangible, not just in the form of a memorial, but via *feelings* associated with them. Thus, places, for Maddrell, are constituted through a "dynamic negotiation of presence and absence" (2013: 504). There is no "static" or essential response to the affects of the past or absent. Instead, one becomes attuned to them, as Maddrell writes:

> I see memorials everywhere and am sure others do too [...] Our landscapes are overlaid with many other "scapes" including deathscapes, of which memorialscapes are part, and these inflect our experience and understanding of those landscapes.
>
> (2013: 210)

These memorials may take less recognisable forms, producing less obvious (yet no less affecting) hauntings. For example, Maddrell considers temporary (and permanent) roadside offerings following fatal traffic incidents. Near Oxford, following a tragic car crash in 2005 in which four people were killed, a new metal barrier to separate lanes of traffic was installed:

> For family, friends and other locals who know of this tragic case [...] the mile-long crash barrier itself evokes the absence-presence of the deceased, and is itself a lasting memorial, the permanent transformation of the material landscape hinting at the radically transformed emotional landscape of the lives of the bereaved.
>
> (Maddrell, 2013: 517)

Importantly, memorials of all kinds often pass unnoticed or are interpreted very differently. Indeed, "one person's sacred memorial can be another's litter or vandalism visited upon the landscape" (Maddrell, 2013: 210).

> ***Pause for thought:*** *Can you think of a time where you encountered an object in the landscape that held little significance for you, but which may have been important to someone else? Alternatively, can you think of an object in a landscape that has some emotional importance for you, but which goes unnoticed for other landscape users?*

From such accounts, we can see how an attention to ghosts might bring forth new experiences, and new questions, about landscape. In this vein, the sociologist Avery Gordon characterises the "unsettling" nature of such haunted spaces. The "ghosts" of the absent past are transformative as they force us to confront the continuing presence of the past, and the violence and injustices involved. They propose a different way of knowing or alternative histories of place (as discussed in Box 11.3). Haunting occurs when one encounters things that affect a sense of the uncanny in a place. Alternatively, such "ghosts" might appear with experiences of grief or reflect the memory work practised through a particular space. Landscapes "are *always* made up of losses, absences and presences – but differently so for different people, and in changing constellations of asymmetrical tension" (Jones, 2015: 11 – orig. emph.). Finally, haunting can manifest in direct encounters with absence as a form of presence: where the fact of absence itself is a constituent part of the meaning of the landscape as the fact that this or that is *not there* has as tangible a force as that which remains.

Spectral Ecologies

Before closing the chapter, we consider the ways in which an analytic of haunting is especially appropriate today: for engaging the "spectral ecologies" characteristic of widespread biodiversity crisis and their manifestation via landscapes of decline or extinction (see Garlick, 2019; Newman, 2025). As discussed elsewhere, the extent of human impacts upon the environment, and the traces left within the stratigraphic record, has seen some characterise our geological epoch as the "Anthropocene", or age of humans (see Chapter 10). Central to these characterisations has been the prevalence of degraded and deteriorating ecologies, themselves often marked or encountered through the absence (or near absence) of particular species, phenomena, or relationships. The claim that we are living through a period of human-driven extinction suggests many landscapes, previously animated or perceived in relation to specific nonhuman presences or processes, today fundamentally changed or changing. In turn, experimental, creative, and technoscientific forms of conservation facilitate the return of "revenant" species by way of reintroductions, rewilding initiatives, and even attempts at cloning (Searle, 2022 – see Box 11.4).

What are we to make of landscapes haunted by such processes? The material, documented, and affective traces of past or absent forms of life suggest a need to consider landscape as abounding with "more-than-human memory" (Newman, 2025: 84). For example, anthropologist Hugo Reinert, drawing upon fieldwork exploring seabird conservation in the Norwegian Arctic, discusses how an experience of silence in the landscape characterises the decline and absence of species, and a sense of loss. Thus, through landscapes as he describes on the island of Vaerøy, cliffs once busy with nesting seabirds are now empty and possibly abandoned, "listening for the sound of nothing" opens onto "a sort of background unease: a nagging and unclear affect [...] a sense of something missing" (Reinert, 2018: 503). The potential extinction facing many seabird species, owing to pressures including plastic pollution and sea-warming impacting food supplies, produces a haunted landscape imbued with an uneasy, troubling silence that provokes "strangeness […] grief, guilt, anger and nostalgia" (Reinert, 2018: 507). In this work, then, direct engagement with a given landscape offers a means of grappling with the proliferation of ghosts arising out of the "ruptured" (after Reinert, 2015) ecologies that (certain) human cultures are producing and leave in their wake.

Box 11.4 Hunting Ghosts in the Pyrenees

The bucardo, or Pyrenean Ibex, is a species of wild goat that has long been prized and mythologised by hunters in the mountainous border region between Spain and France due to its elusive existence, and the skill required to track and kill one. Across his research, Adam Searle (2021, 2022) has examined how the spectral, ghostly nature of the bucardo haunts and animates the cultural landscape of the Pyrenees, imbuing it with a spectacular sensibility, despite its apparent total absence following species extinction upon the death of the final individual in 2000.

The earliest accounts of the species' existence were in the context of a medieval treatise on hunting, as Searle describes. Hunters were drawn to places such as the Ordesa Valley in the Aragonese Pyrenees attempting to take and display a bucardo trophy as a "prized ornament […] with a romantic story" of human vanquishing wilderness (2021: 514). Long considered the "top prize" because of the way its elusiveness presented a challenge testing the wits of any pursuer, the association between the Ibex and the landscape informed the latter's sense as a challenging wilderness, accommodating the former as a "spectral spectacle" that comprised, in nineteenth-century accounts, "an intractable aspect of the Pyrenean Landscape, difficult to kill, a spectre" (Searle, 2021: 522). Despite efforts at conservation, the species approached extinction for much of the twentieth century. As Searle discusses, the association of the bucardo with extinction meant that traces of its presence or material artefacts from animals shot within the Pyrenees evoke and impress a profound sense of absence bound up with what this environment means and how it feels. The trophies of hunted animals are themselves spectral objects, sitting out of time and place, material reminders of the absence of *living* bucardo. Hunters' accounts and testimonies, alongside images and photographs of the animal, recall its absences yet ensure its continued association.

Since 2014, translocated Iberian Ibex reintroduced as "ecological and cultural proxies" evoke, in another way, the spectral return. These animals are not bucardo, yet via reintroductions, breeding, and adaptation to a new environment, they stand in for, and become increasingly indistinguishable from, them. The meeting of the physical landscape and animals' bodies is producing a creature with a physiognomy strikingly reminiscent of the extinct bucardo. In this context, Searle's work asks, what, when a species returns from extinction, remains lost from the cultural landscape (see Garlick, 2019)? Are claims of enduring absence in the face of this returned presence significant, or are the differences between bucardo and Iberian Ibex "fabricated for political ends" (2022: 176) by different interest groups merely to support different ways of doing conservation or land management? The ways in which the species haunts the Pyrenees and both their imaginary and actual ecology unsettle these mountains (Searle, 2021: 525).

Pause for thought: *Can you think of another landscape that is associated with the absence of a particular lifeform? Does one require knowledge, training, or experience to become aware of that absence, or is it immediately recognisable? How does acknowledging this absence change one's experience of the landscape in question?*

Chapter Summary

Box 11.5 Key Concepts

Chapter 11

Uncanny
Weird and Eerie
Absence
Ghosts
Haunting

In this chapter we have considered how studies of cultural landscapes, drawing on contributions from non-representational theory, might attend to hauntings and ghosts (Box 11.5). These terms have offered a means to consider both landscape's "uncanny" nature in its weirdness and eeriness; and its capacity to surprise, destabilise, unsettle, and undermine clear borders or boundaries. In these ways, landscapes propose that we might yet get lost within them.

In contrast to previous chapters, our concern here is with landscapes textured by absences, rather than presences. That is to say: how the things not there, or not available, within cultural landscapes return or exert a force upon us, perhaps by way of material traces (such as memorials) or in terms of the emotional knowledge associated with places. Recognising cultural landscapes as haunted acknowledges the influence of pasts and absences, conceptualised by means of hauntings and ghosts, and their continuing affective "push" on our experience. The figure of the ghost captures the myriad ways in which the past bleeds into the present with affective force.

Finally, we have considered the ways in which the past can haunt us in surprising ways. Some "ghosts" require us to tune in to landscapes differently; others reveal alternative, possible versions of landscape and prompt interesting questions about their character and formation. The politics of how ghosts are made present, seen, engaged, or mobilised in support of contemporary cultural agendas, narratives, and interpretations reveals the significance of absence in shaping human–environmental interactions and relationships.

Suggested Seminar Activities

Activity 1: Discuss one of the examples introduced in this chapter or located elsewhere in the recommended reading.
For the chosen example consider the following:

1. How does this particular landscape prompt you to move beyond thinking of its geographies merely in terms of *presence* (i.e. bodily presence, encounters with things there)?
2. How might the conceptual framework of haunting help account for this landscape's working?
3. If you were to consider this landscape as a "haunted place", what is it specifically about this landscape or its formation that terms like "haunting" or "ghosts" direct attention towards?
4. With regard to the example landscape, how might encounters with its ghostly elements prove "*transformative*" (personally, socially, culturally, emotionally, politically, or otherwise …) or come to be used in ways that are *political*?

Activity 2: Engage in a "free writing" exercise. Use 2 minutes to write *continuously* in response to one of the following prompts. You do not have to share this piece of writing, but the ideas/general content should then serve as a prompt for further discussion in groups.
Themes:

- Write about a time you experienced a landscape as weird, eerie, creepy, or spooky.
- Write about a landscape associated with a particular memory from your childhood.
- Write about a landscape that you associate with specific historic events.

Next discuss the general content, ideas, and themes that emerged from before considering the following questions:

1. Can you remember how you perceived this landscape differently to how you do now?
2. How are memories provoked by being in the landscape (e.g. under particular conditions, whilst in particular emotional state)?

3 Was this experience shared with anyone else, is this memory/experience of landscape made more present in company?
4 Do you associate this experience of landscape with any particular bodily sensations or feelings?

Recommended Reading

Bordelon, J. 2022. 'I dreamt about it, just like this, exactly the way it is': Haunting emotional geographies in Buenos Aires. *Emotion, Space and Society* 43: 100881.
Jones, J.P. 2019. What's the matter with ghost stories? *Geohumanities* 5(1): 278–281.
Reinert, H. 2018. The haunting cliffs: Some notes on silence. *Parallax* 24(4): 501–512.

References

BBC News. 17 May 2006. Piano buried on UK's highest peak. Accessed online at: http://news.bbc.co.uk/1/hi/scotland/highlands_and_islands/4990662.stm on 22 May 2025.
BBC News. 2 December 2016. Mystery 'banging' noise solved in Queens Park, Bolton. Accessed online at: www.bbc.co.uk/news/uk-england-manchester-38180456 on 22 May 2025.
Bordelon, J. 2022. 'I dreamt about it, just like this, exactly the way it is': Haunting emotional geographies in Buenos Aires. *Emotion, Space and Society* 43: 100881.
Cambridge Dictionary Online. 2024. "Uncanny". Accessed online at: https://dictionary.cambridge.org/dictionary/english/uncanny on 16 August 2024.
Dekeyser, T., Secor, A., Rose, M., Bissell, D., Zhang, V., and Romanillos, J.L. 2022. Negativity: Space, politics and affects. *Cultural Geographies* 29(1): 5–21.
Derrida, J. 1994. *Spectres of Marx*. London: Routledge.
Fisher, M. 2011. *Ghosts of My Life: Writings on Depression, Hauntology and Lost Futures*. London: Zero Books.
Fisher, M. 2016. *The Weird and the Eerie*. London: Repeater Books.
Freud, S. 1919/2003. *The Uncanny*. London: Penguin.
Garlick, B. 2019. Cultural geographies of extinction: Animal culture among Scottish ospreys. *Transactions of the Institute of British Geographers* 44(2): 226–241.
Garlick, B. 2023. The total mountain: Nan shepherd and the virtual qualities of landscape. In: Hall, M. and Hall, J. Eds. *The Mountain and the Politics of Representation*. Liverpool: University of Liverpool Press, pp. 211–231.
Gordon, A. 2008. *Ghostly Matters: Haunting and the Sociological Imagination*. Minneapolis, MN: Minnesota Press.
Jones, J.P. 2019. What's the matter with ghost stories? *Geohumanities* 5(1): 278–281.
Jones, O. 2015. "Not Promising a Landfall ...": An autotopographical account of loss of place, memory and landscape. *Environmental Humanities* 6: 1–27.
Kemmer, L. and Simone, A. 2021. Standing by the promise: Acts of anticipation in Rio and Jakarta. *Environment and Planning D: Society and Space* 39(4): 573–589.
Kingsbury, P. and Secor, A. 2021. *A Place More Void*. Lincoln, NE: University of Nebraska Press.
Latour, B. 2005. *Reassembling the Social: An Introduction to Actor-Network Theory*. Oxford: Oxford University Press.
Luckhurst, R. 2002. The contemporary London Gothic and the limits of the 'spectral turn'. *Textual Practice* 16(3): 527–546.

Maddrell, A. 2009. Mapping changing shades of grief and consolation in the historical landscape of St Patrick's isle, isle of man. In: Smith, M., Davidson, J., Cameron, L., and Bondi, L. Eds. *Emotion, Place and Culture.* Farnham: Ashgate, pp. 35–55.

Maddrell, A. 2013. Living with the deceased: Absence, presence and absence-presence. *Cultural Geographies* 20(4): 501–522.

Mori, M., MacDorman, K.F., and Kageki, N. 2012. The uncanny valley [from the field]. *IEEE Robotics & Automation Magazine* 19(2): 98–100.

Morton, L. 2015. *Ghosts: A Haunted History*. London: Reaktion.

Nancy, J-L. 2005. *The Ground of the Image*. Bronx, NY: Fordham University Press.

Newman, M. 2025. Remembered belonging: Encounters with the spectral more-than amidst landscapes of decline. *Cultural Geographies* 32(1): 75–91.

Paphitis, T. 2020. Haunted landscapes: Place, past and presence. *Time and Mind* 13(4): 341–349.

Pile, S. 2005. *Real Cities*. London: Sage.

Reinert, H. 2015. The landsacpe concept as rupture – Extinction and perspective in a Norwegian Fjord. In: Sooväli-Sepping, H., Reinert, H., and Miles-Watson, J. Eds. *Ruptured Landscape: Landscape, Identity, and Social Change.* London: Springer, pp. 41–54.

Reinert, H. 2018. The haunting cliffs: Some notes on silence. *Parallax* 24(4): 501–512.

Rose, M. 2010. Back to back: a response to" Landscape, absence and the geographies of love". *Transactions of the Institute of British Geographers* 35(1): 141-144.

Royle, N. 2003. *The Uncanny*. Manchester: Manchester University Press.

Searle, A. 2021. Hunting ghosts: On spectacles of spectrality and the trophy animal. *Cultural Geographies* 28(3): 513–530.

Searle, A. 2022. Spectral ecologies: De/extinction in the Pyrenees. *Transactions of the Institute of British Geographers* 47: 167–183.

Thrift, N. 2008. *Non-Representational Theory: Space, Politics, Affect*. London: Routledge.

Till, K. 2005. *The New Berlin: Memory, Politics, Place*. Minneapolis: University of Minnesota Press.

Tolia-Kelly, D.P. 2016. *Landscape, Race and Memory: Material Ecologies of Citizenship.* Abingdon: Routledge.

Wylie, J. 2009. Landscape, absence and the geographies of love. *Transactions of the Institute of British Geographers* 34(3): 275–289.

Wylie, J. 2016. A landscape cannot be a homeland. *Landscape Research* 41(4): 408–416.

12 Therapeutic Landscapes

Introducing Therapeutic Landscapes

In her 2017 memoir, *Swimming with Seals*, Victoria Whitworth offers a neat example of how particular places and activities can make us feel better. Using the practice of cold-water swimming, she illuminates the relationship between health and landscape by articulating the ways in which the sea helped her to deal with experiences of crisis and loss:

> Swimming in the sea has helped me rediscover myself. It has opened Orkney up for me, and made me understand the strengths of my body, as well as its limits. It has been fundamental in healing my feet, but, more than that, it has allowed me to fall in love with being embodied all over again. With being alive.
>
> (Whitworth, 2017: 274)

While dousing oneself in icy waters might not strike everyone as immediately therapeutic, wild swimming has become more popular in recent years (Figure 12.1). Articles, books, and podcasts all extol the virtues of a cold-water dip. *The Guardian* (Foote, 2025) describes the experience as "the ultimate tonic", *Australian Vogue* (Day, 2 January 2020) as "one of the best things for your mental health", and even the *British Medical Journal* endorsed swimming outdoors as a treatment for major depressive disorder (Van Tulleken et al, 2018). Groups like the "Polar Bear Club" have been set up by avid swimmers to demonstrate dedication, with members required to swim 200 m outdoors twice a month between November and March wearing only a swimsuit, goggles, and rubber hat.

The idea of entering a given landscape to experience health benefits is not new. From the range of Victorian spa and seaside resort towns in the UK, like Scarborough; to the ancient Sanctuary of Asklepios in Greece (today a UNESCO site of Outstanding Universal Value), many societies have long envisaged a relationship between particular places, activities, and health. It was not until the 1990s, however, that cultural landscape studies began fully engaging with this question, and the concept of the "therapeutic landscape" was outlined in earnest. It is to this concept the chapter turns, highlighting not only important traditions within its

DOI: 10.4324/9781003172451-13

Figure 12.1 Wild swimming in the Cairngorms, Scotland, UK, in 2023. Image by Rachel Hunt.

associated approaches, but also the nuances, critiques, and potential applications of such landscape research.

To explore therapeutic landscapes, this chapter takes a socio-ecological approach to health. In other words, it understands health as shaped by and through a multitude of interacting personal and environmental factors. Health is viewed as more than what occurs within the body to also encompass the relationship between that body and its wider environment. Thus, the term "therapeutic landscapes" condenses the idea that particular landscapes – or, as examined below, *experiences in* landscape – can have a positive influence upon the health and wellbeing (physical and/or mental) of individuals, groups, or society at large. In this way, researchers on therapeutic landscapes' engagement with landscape theory mirror, to a certain extent, the developments considered over Chapters 2–10 as the environment is variously conceived as viewed or "read", experienced, felt, and related to.

Chapter Outline

This chapter establishes foundations for scholarly interest in the relationship between cultural landscapes and health. It builds on previous chapters to chart how landscapes have come to be appreciated as therapeutic in their cultural, symbolic, and phenomenological dimensions. Exploring the origins of such interest in health and landscape, we discuss the focus on "non-ordinary" environments, greenspace, and nature, alongside the rise of "colour-based" approaches to conceptualising environments' therapeutic qualities. Outlining a range of key critical debates with these ideas as they raise questions of subjectivity, relationality, and "palettic" approaches to health and space, the chapter then concludes by reflecting on how "therapeutic landscape" scholarship has been applied beyond the academy in the realms of policy.

Connecting Health with Landscape

The notion of therapeutic landscapes was significantly outlined by American geographer Wilbert Gesler in 1992. Prompted by the appearance of a "new" cultural geography, with a basis in cultural studies that developed the landscape concept beyond what was seen as a largely descriptive and a-critical framing (see Chapters 1–2), Gesler felt such ideas might likewise expand understanding of the relationship between health and the environment. In particular, key concepts proposed or taken forward within cultural landscape literature (including "sense of place", "hegemony", "landscape as text", or "symbolic landscape") could also "be applied to the explication of therapeutic processes in various settings" (Gesler, 1992: 735) and complicate the way in which "medical geography" had tended to appreciate this connection (see Box 12.1).

Box 12.1 Medical Geography

Some form of "medical geography" has been present within writing and research for centuries. In essence, it concerns the study of the relationship between health (or sickness) and the environment to determine how one influences the other. In scientific terms, we might consider contemporary epidemiology, and its interest in the spread and distribution of disease, as a version of "medical geography". In terms of social sciences, a historic body of commentary has framed and examined the relationship between human activities, environments, and ill-health, often with the intention of influencing and informing policy or management.

An early example of medical geography might be seen in the writing of ancient Greek philosopher Hippocrates, who "stated the importance of 'airs, waters, places' as an influence on human health, achievements and history" (Philo, 2009: 451). We might also understand miasmatic theories of disease that associated sickness with "bad air", and the environmentally deterministic discourse of "tropicality" as central aspects of the medical geography informing colonial encounters with the landscape in places such as India, Singapore, or Sri Lanka (see Arnold, 1993). More recently "health geography" (see Kearns and Moon, 2002) understands health as more complex than the absence of disease. Amongst other foci, current work emphasises the relationship between healthy subjects and places, spaces, environments, and landscape. In addition, a broader and interdisciplinary interest in medicine, health, and wellbeing that cuts across a number of subjects, gathered under the heading of the "Medical Humanities" or "GeoHumanities", demonstrates how the topic of health (and associated questions) has become central to the work of many within subjects traditionally not associated with such questions (see Atkinson and Hunt, 2022).

With an expanded understanding of landscape informed by the cultural turn, Gesler's work throughout the 1990s examined key sites possessing historical or contemporary reputations for healing and health. In notable examples, he considered the Marian Shrine at Lourdes, France (Gesler, 1996) and the Roman baths in Bath, England (Gesler, 1998), as "therapeutic landscapes". That is, as tangible environments with contextual and perceived cultural or material relationships to physical or mental health.

Key Components of the Therapeutic Landscape

In Gesler's studies of landscape, the "therapeutic landscape" was both a conceptualisation – or theory – of landscape and a particular kind of environment. He examined "places that have achieved lasting reputations for physical, mental, and spiritual healing" (Kearns and Gesler, 1998: 8). Importantly, within his work, both material and *social* dimensions of the landscape impacted health. How the landscape was organised, engaged with, or thought about could be as "therapeutic" as its environmental characteristics. A therapeutic landscapes approach therefore departed from the premise that human mind and material world combined created landscapes beneficial to health.

Considering landscapes known culturally to be associated with therapeutic properties, Gesler (1993) proposed each be examined in terms of the four, interacting aspects crucial to their achieving a lasting reputation as sites of healing. Specifically, he examined relationships between, firstly, the "natural" (understood as the external, nonhuman world); secondly, the built environment (structures made by humans); thirdly, the symbolic (how specific meanings were associated with or inscribed onto the environment); and, fourthly, the social (how places were lived in and experienced by human society). This four-way typology underlays his conceptualising of how therapeutic landscapes operated and is further expanded in Box 12.2.

Box 12.2 Gesler's Typology

William Gesler's four-way typology was first deployed in discussion of the healing temple of Epidauros, Greece. He examined how the association between this site and its therapeutic effects operated via four key registers: the natural, built, symbolic, and social.

The Natural: Natural places have been thought for thousands of years to have curative powers, whether scientifically proven or not. There is a long history of going into nature for therapeutic means and blue, watery spaces have also long held similar connotations.

The Built Environment: The built environment can be shaped to create specific moods, often in combination with the natural. Gesler later used the example of "a hospital serving the Māori in a remote area of [Aotearoa] New Zealand sited on a spot that provides patients with striking views" (1993: 174).

The Symbolic: The importance of symbols in creating landscapes (note here links to Cosgrove, 1987) can be seen in both concrete symbols and also the way that physicians hold an elevated, almost sacred status. This can also be seen in language through the way in which mental illness has historically been classified as both "deviant" and a "threat to society" (Clark and Dear, 2020 [1984]: 74).

The Social: In Gesler's last category, the social, we can also see evidence of the phenomenological notion of landscape, as something produced through experience as well as vision. In this category the patient, and their social environment, is integral to the process of healing.

Pause for thought: *Consider a landscape you associate with benefits to your physical or mental wellbeing. Consider how Gesler's typology might be applied to this landscape, and the different elements of it that might be grouped within each of the four categories.*

Whilst this attention to therapeutic landscapes emerged primarily within health geography, it also exists in conversation with ideas examined elsewhere in this textbook. In the introduction to the collection *Putting Health into Place: Landscape, Identity, and Well-Being* (1993), Gesler, alongside co-author Robin Kearns, explicitly notes the concept of "therapeutic landscape" owes much to cultural landscape theory. Indeed, it can be seen to have emerged out of the work of the geographical "humanist" tradition, as well as a broader "phenomenological" approach to environmental knowledge (see Chapter 8). This scholarship attended to the textures, sights, sounds, and smells that "create a subtly unique 'feel' to places that are of immense importance to life" (Milligan et al, 2004: 1783). As humanist geographer Yi-Fu Tuan notes, "[l]andscape is a construct of the mind and feeling" (1979: 89).

It is also "sticky" with affect and emotion (see Chapter 9), stimulating or prompting us to react in unexpected or surprising ways. In terms of emotion, Tuan's conceptualising of "place love" ("topophilia") and place aversion ("topophobia") can clearly be connected to a sense of our being drawn towards (or away from) sites understood as "therapeutic".

The "Non-Ordinary" Landscape: Green, Wild, and Watery

Early efforts to follow Gelser and examine a wider range of therapeutic landscapes originally tended to focus on "non-ordinary" spaces. These were sites including temples, spas, or health retreats, as well as green or wild spaces and their use or experiencing via vulnerable groups. We examine some of these strands of therapeutic landscape study below.

Greenspace and Nature in Therapeutic Landscapes

The association of greenspaces with health is well documented. This includes discussion of the health benefits of "green space" (e.g. parks, forests, nature reserves) and the conceptualising of "greenspace" as a way of articulating that relationship. As geographer Hester Parr (2007) has described, mental health conditions in the eighteenth and nineteenth centuries were understood by some doctors of the time as evidence of too little contact with "nature"; a consequence of increasingly urban-centred, industrial societies. As a curative, patients were effectively removed to rural landscapes, whose "pleasant order" was thought to soothe and promote "calm, rational reflection" (Parr, 2007: 540). Such landscapes were hardly wild, untamed, or dangerous (as those discussed below), but manicured, designed, and controlled. As Parr writes, "nature, here, was seen as a therapeutic space, the promise of health premised in part on the passive absorption of the inherent healthy and rational properties of moral and manicured asylum land" (2007: 541). Such treatments in historic Britain exemplify the long Western tradition of "retreating to nature" for self-improvement (Conradson, 2016: 104). As discussed in previous chapters, such narratives are premised upon a specific discourse of nature as separate and external that persist in contemporary culture (see Box 12.3).

Work in the vein of outlining the health benefits of greener landscapes relied upon a widely assumed, generally agreed correlation between greenery and "emotional gains" (Conradson, 2005a: 103). Such landscapes were those "in which a connection can be made with nature in a return to a premodern engagement with 'pure' nature" (Little, 2015: 799). Researchers therefore sought to explore how such landscapes "have the capacity to shape our feelings" (Conradson, 2005a: 107) or even conjure specific emotions (Ewert et al, 2003), resonating with the notion of emotional landscapes or geographies (see Chapter 9). The idealised benefits of greenspace, or what Foley (2018) terms the "*vitamin G*" effect, have been showcased across various settings and scales from National Parks to local playgrounds; fields to expansive woodlands, and across urban landscapes including gardens and allotments (see below).

Box 12.3 Therapeutic Wilds in Fiction and Film

The notion of a "retreat to nature" often evokes a seeking out of "wilderness". The framing of wilderness as restorative and nourishing owes much to the traditions of eighteenth- and nineteenth-century Romantic poetry, literature, and art and today we can see its continuing currency within popular cinema.

For example, the film *Wild* (2014) capitalises on an idea of wilderness as a space for both the renewal and reinvention of the self. The film itself is based on the memoir *Wild: From Lost to Found on the Pacific Crest Trail*, by American author, Cheryl Strayed. Following her mother's death, the breakdown of her marriage, and a period of drug addiction, Strayed decided to walk the 1,000-mile-long Pacific Crest Trail. Her account, adapted for the screen, was a box-office sensation starring American actor Reese Witherspoon and encodes the wilderness landscape, located beyond the limits of human habitation or control, as a site of renewal. A similar narrative is at work within the film *Leave No Trace* (2018) by filmmaker Debra Granik, which concerns a troubled war veteran, afflicted by PTSD, attempting to live with his daughter in the forests of Oregon, US. His efforts to live within society see his mental health deteriorate and he (and his daughter) come to realise that only in "the wild" can he truly find some form of peace.

Notions of nature and "natural" landscapes as restorative are therefore persistent and widespread within popular imaginaries of the "therapeutic landscape". In different ways, the films above package wilderness as a space for mental or physical regeneration. Such a vision of wildness is famously conveyed by Henry Thoreau in his transcendentalist account *Walden* (1995 [1854]), detailing two years spent living in relative isolation in the woodlands of Massachusetts in search of Enlightenment through immersion in nature. Forcing individuals to live differently, and consider themselves anew – by confronting them with death, instilling confidence in their abilities, and taking them towards some kind of spiritual reckoning – the opportunity arises for the recovery of a "natural" state of wellbeing denied by modernity (Vannini and Vannini, 2016). Such associations with the wild have a long history in Western thought and, especially, American narratives of the "frontier" (see Cronon, 1996; and Chapter 2). A dissatisfaction with materialist, industrial society, and belief that intimate contact with nature and solitude are beneficial, thus frames wilderness as "not unlike an island: a place seemingly disconnected from the rest of the world where one can travel to break ties from the past and reinvent oneself from scratch" (Vannini and Vannini, 2016: 77). Whether therapeutic exposure comes in the form of wilderness, a garden or a forest, as Foley states, "the ongoing power of green spaces to act as sites of rest and restoration, within which stillness, escape and shift-spaces from everyday stress and pressure [have] remained important tropes" (2018: 253).

Palettic Approaches: Blue, Brown, and Blurred Space

While greenspace is prominent within studies of therapeutic landscapes, environments besides those categorised as "green" or "wild" also feature in discussion. From inception, therapeutic landscape scholars have acknowledged the benefits of water. Indeed, it was via interest in watery spaces that Gesler developed his account of healing places as culturally (re)produced through their reputations, for example by studying the development of spa towns around mineral springs (Gesler, 1992). In turn, recent scholarship has focused on an expanding range of "blue" spaces. This term encompasses "a range of natural and artificial spaces associated with water, including lakes, rivers, seas, reservoirs and other pooled waters" (Foley, 2018: 251).

Some scholars have focused on the specific therapeutic qualities of water itself, proposing terms such as "healthy blue space" (Foley and Kistemann, 2015). Here the emphasis is upon landscapes with the potential to be health-enabling because of their watery properties, including coasts, lakes, rivers, waterfalls, or even lidos. Through this lens, practices such as wild swimming (re-)appear as cultural landscap(ing) activities which contribute both to the health and wellbeing of those swimming and create "resilient places that continue to provide an ongoing public good" (Foley, 2017: 50). Attention to blue spaces has also encompassed the "urban blue" (Völker and Kistemann, 2015) and the canals, ponds, fountains, or splash pads variously found in cities across the world. In such contexts, the social relations mediating access to particular sites, experiences, or practices (like outdoor swimming) reveal the complexity of the therapeutic and emotional landscape wherein belonging and conviviality of some exists alongside the absence and exclusion felt by certain others (Bates and Moles, 2024).

Collectively, focus on different "colours" of landscape has led therapeutic landscape scholarship to coalesce around particular "palettes of place" (Foley, 2017). Whilst discourse on green and blue spaces is most prominent in academic and policy circles, there is increased acceptance of more diverse "palettes". These include mixtures of "grey" and "brown" space within urban landscapes, "yellow" desert spaces (Wang et al, 2018), and "white" snow-covered spaces. Regarding the latter, Kaelan Brooke and Allison Williams (2021: 1276) propose Iceland as a "white wilderness [...] for wellbeing". As they write, whilst Iceland's landscapes are "not traditionally associated with well-being", given that many in "northern climates often experience snow and ice covered spaces for four or more months per year", it is therefore of value to examine the "therapeutic qualities" of such "white landscapes" as they are mediated by guides and tourist agencies and prompt experiences of awe and wonder (including in relation to "black" lava landscapes).

As the range of colours considered therapeutic is broadened, so is increasing acknowledgement of a blurred spatial palette. Foley's (2015) discussion of the significance of land/water interfaces considers "blue-green" spaces, and the nuance of "brightness" (or shade) in discussing river walks and urban landscapes' therapeutic impacts have also been interrogated (Völker and Kistemann, 2015). Although such engagements with the "palettic" (Bell et al, 2018:124) reflect an appreciation of therapeutic landscapes as complex and multifaceted, there remain limits to the approach's ability to grasp the full spectrum of the environment, not to mention geographical (largely Western) limits concerning its practical application.

> ***Pause for thought:*** *Think of an experience of or encounter with "nature" that you would consider to be restorative. What colours would you (or others) associate with the kind of space in which this experience took place?*

Critical Engagements with Therapeutic Landscapes

Since its early conceptualisation in the work of Gesler and others, notions of the therapeutic landscape have been expanded to consider a range of environments and places including spaces for marginal populations; allotments for refugees (Biglin, 2020), community-based alcohol recovery programmes (Wilton and DeVerteuil, 2006), and digital therapeutic landscapes (Thompson, 2021). There has also been some examination of traditional healthcare settings (e.g. hospitals) via this framework (Gesler et al, 2004). Amidst these developments, scholars have also emphasised that landscape is not the inanimate backdrop for health, but central to its *practised* manifestation. For instance, Jennifer Lea (2008: 91), researching experiences of yoga and massage retreats in Spain, observes that, "being in nature can have a direct therapeutic effect" on the body, it is also vital to examine "how nature is actively enrolled in the creation of therapeutic effect". The key point, then, is that the therapeutic landscape is not merely a *representation* or a set of cultural associations with the environment, but a question of how diverse bodies interact with worlds to produce particular outcomes (see Chapters 8–10).

Acknowledging therapeutic landscapes as practised and performative also prompts several critiques of how the relationship between the environment and health has been conceived. We emphasise four key areas of debate below, encompassing, firstly, the relationality of therapeutic landscapes; secondly, their intersectional character; thirdly, an interest in more everyday manifestations; and, fourthly, challenges to the ways in which "colour" has been applied after the palettic approach. This list is not exhaustive but gives a useful sense of where work in this vein has continued to develop and generate dialogue.

Therapeutic Landscapes as Relational

Geographer David Conradson's work is a good example of how scholars have challenged the assumption of an intrinsic or essentially therapeutic quality being attributed to certain landscapes. Rather, landscapes' therapeutic dimensions are framed in his work as being fundamentally *relational*. In his work on rural English environments, for instance, Conradson argues that an individual's experience of a therapeutic landscape "emerges through a complex set of transactions between a person and their broader socio-environmental setting" (2005b: 338). Thus, a distinction is made between a "therapeutic landscape" and "therapeutic landscape experiences", wherein the latter names a relational outcome involving specific encounters with and responses to a landscape. Focusing upon the self-landscape encounter, Conradson (2005b: 340) draws upon psychotherapeutic understandings of the self, whereby the human is "not [...] an autonomous, tightly bounded entity but rather [...] something that emerges within and through its relations to other people and events" (see also Chapter 10).

Such understanding of therapeutic experiences also connects with landscape phenomenology. The self, like the landscape, is a product of its interaction as people "fold particular events into their selves" (Conradson, 2005b: 340). Interviewing guests and staff at a respite centre in Dorset, UK, Conradson explores the emergence of a particular "therapeutic experience", describing its production via the relationships involving bodies and place. For example, taking oneself away from the normal home space provides a simultaneous distancing from domestic tasks and expectations thereby creating the sense of a retreat space as therapeutic (it is defined in relation to another space, and an individual's experience within such spaces). The recognition of such a "relational kernel" (Bell et al, 2018: 123) has come to frame much therapeutic landscapes scholarship, including work on spaces of phobia (see Davidson and Parr, 2007), "therapeutic taskscapes" (Dunkley, 2009), and "therapeutic mobilities" (Gatrell, 2013). All of these examples underline the importance of recognising how landscapes will mean different things at different times to different people under different circumstances.

Therapeutic Landscapes as Intersectional

The ways in which individuals or groups might be less able to access therapeutic experiences within particular contexts has also received scholarly attention. These inequalities reflect the personal or social histories as they affect a particular individual or group. Health researchers Christine Milligan and Amanda Bingley (2007) observe, for instance, a correlation between time spent in woodland during childhood and a later capacity to make restorative connections with such landscapes. However, it is not only previous experience that shapes such outcomes. An interest in the *intersectionality* of therapeutic landscapes – that is, an appreciation of how various cross-cutting elements of one's identity shape and inform the self, as well as texture their experiences (such as of marginalisation or privilege) – has also been advocated. Alongside the influence of race or racial politics, for example as it mediates an experience of the English countryside (see Chapter 5), so too can economic disadvantage (Ward-Thompson et al, 2010) or (dis)ability impact a person's capacity to access or make realisable the therapeutic landscape benefits of landscape (e.g. see, Bell, 2018 on the impact of visual impairment). Uncertainty and difference in the therapeutic experience is thus a product of intricate, embodied encounters that unfold within a world shaped and lived through all manner of power relations. Therefore, "understanding how health and place are intertwined remains a crucial task" for addressing such inequalities (Conradson, 2014: 611).

Beyond "Special" and "Natural"

Critique has also been forthcoming concerning the kinds of sites deemed worthy of study. Cameron Duff (2011) has highlighted the enduring focus on "natural" spaces that has long characterised therapeutic landscapes scholarship. In turn, everyday, urban, or so-called third spaces and landscapes have been historically overlooked. As a substantial portion of therapeutic research has turned to examine comprising what can be called "nearby nature" (Foley, 2018: 255), there have been further moves to incorporate other vernacular and working environments (such as the perception of therapeutic landscapes as experienced by urban domestic workers during Covid – see Winata and McLafferty, 2023). Thus, it is recognised that "healing [or ill-health] can take place in everyday, ordinary places, whether a residential backyard, a hospital room or an imagined landscape" (Williams, 2007: 2). As work by geographer Niamh Shortt et al (2017), detailing the variety of environments integral to alcohol recovery, makes clear, just as any space can be therapeutic so too can such spaces change between (or for) individuals depending on day, time, or other influences. Moreover, in the context of the intersectional work discussed above, one should not presume that "nature" is automatically the location for all activities deemed therapeutic or conducive to wellbeing.

Beyond "Palettic" Approaches

In 2019, Sarah Bell and co-authors summarised the emerging critique of "palettic" approaches to therapeutic landscapes and the tendency, during the prior decade, to colour-code environments. As they observe, the approach had granted too much authority to the visual perception of landscape. Moreover, "framing the environment primarily through colour" risked "overlooking the wider textures, terrains, auditory tones, smells and sensations that are felt through the body to render such encounters therapeutic or otherwise". For Bell and her colleagues, the palettic required reappraising in broader terms, with "a shift towards palettic 'sensescapes'" proposed to "better equip researchers to engage with corporeal contingency and difference" (Bell et al, 2018 124–125). That is, the ways in which different green, blue, brown, or otherwise coded spaces might be differently experienced, sensed, and known via the full gamut of bodily capacities.

The areas of critique discussed above highlight that landscapes are not inherently therapeutic. Rather, we need to ask of them: therapeutic for whom, when, and under what conditions? Such questions engage the politics of therapeutic landscapes, and the (in)equality of their provision or experience. The therapeutic experience is subjective and relational (dependant on who you are, what you have experienced before). In particular, and in response to such a recognition, research on therapeutic landscapes has begun to consider in more detail the role of the senses. It is therefore important that understanding moves away from an *a priori* definition of what therapeutic landscapes are (after Gesler's early work), to instead focus on the individual responses which shape more relational, less predictable experiences (see Box 12.4).

Box 12.4 "Cultivating Health" in Garden Spaces

Christine Milligan et al's (2004) work with older people and Josephine Biglin's research with refugees (2020) have both focused on the everyday space of the allotment garden to explore the therapeutic landscape as relational. Both of these studies suggest that gardening activity is one way to tackle loneliness in vulnerable groups; and both explain this in a way which also acknowledges that health promoting environments can be actively created with different benefits depending on those who use them (in this instance, they focus on older people and refugees respectively).

These benefits are accrued partially through passive enjoyment of the garden (e.g. "seeing the acorns and the different seasons" (Milligan et al, 2004: 1785) but also via the social side of activities, which, as Biglin highlights, can be "meaningful, irrespective of verbal communication" (2020: 4). In the case of Biglin's work, such social interactions alleviated some of the pressures of the refugee experience, particularly with respect to verbal communication, whereby immersion in the practical activities of the allotment (as a therapeutic taskscape) altered expectations around discussion or chat. In the words of one of Biglin's participants, "I am still learning English so I don't always understand everyone but it doesn't matter so much because we are working" (2020:4).

Understanding participants and their experiences as intersubjective entails also recognising that they bring with them a host of past knowledge and experience, which contributes to how a place holds meaning in the present. For the participants of Biglin's work, a familiar scent or sound, a plant of the same colour, or an activity (gardening) which held significance in a previous life can add a "*bit of home*" and be "*really important*" (Biglin, 2020: 5). Qualitative research such as this offers an acute reminder that therapeutic landscapes are complex, sometimes craft-able and shared, but always diverse and subjective.

Therapeutic Landscape in Application

From its inception, the therapeutic landscape concept has been consistently applied in a way that emphasises its potential policy relevance. Gesler argued in 1993 that it was "important to analyse in some depth the meaning and context in which healing is provided in specific times and places" (1993: 184) such that they could be managed for the benefit of users. In the early 2000s (and as discussed in Box 12.4), Milligan et al's (2004: 1784) work on allotments would emphasise how support for communal gardening sites could provide "one practical way in which it may be possible to develop a more 'therapeutic landscape'" for older individuals, alongside the need to expand opportunities for developing and engaging with such environments in urban areas.

A challenge for therapeutic landscape researchers seeking to influence policy and landscape planning is thus not only to identify and celebrate the attributes of therapeutic spaces (such that they might be cultivated), but also to broaden the concept of health (mental and physical) as relationally and subjectively bound up with a diversity of (urban and rural) environments. There is need to create spaces fostering social connectedness and happiness as much as those easing traffic congestion or improving air quality. Each, however, concerns different kinds of landscape, and different ways of *knowing* (measuring or researching) landscape.

Therapeutic landscapes research is important especially within applied disciplines including planning, public health, and landscape architecture. Across such fields, similar terminology (e.g. "green", "blue", and "brown" space) and methodologies underline an interconnectedness of language and approach. Indeed, palettic notions of the therapeutic landscapes have been taken up by several government bodies, such as in the UK, and enshrined within policy (see Box 12.5) such that they now have a tangible impact on how landscapes are practically understood and organised.

Box 12.5 Therapeutic Landscapes in Policy

The notion of colour-coded spaces associated with specific therapeutic benefits has become enshrined in public policy in Britain, as evidenced by a range of examples. In 2016, the UK governmental organisation Public Health England stated:

> We've known for some time that good quality natural landscape in urban areas can affect how people feel. It reduces stress and sadness, lifts the mood and makes us feel better. The benefits of these green and blues space, and the mechanisms by which they work, are varied. Some are the physical benefits from green infrastructure, for example improved air quality, less noise pollution and reduced risks from flooding or heatwaves. There are also benefits to active users of these spaces, whether that's physical recreation or through children interacting with nature. The impacts on mental wellbeing, social networks and sustainable communities probably work through a variety of mechanisms and it's in these areas that the strongest evidence is emerging that urban green space can improve the public's health. We know people who live in the areas within our cities and towns that have more green or blue space have better mental health.
>
> (Morton, 2016)

Such a statement deploys the therapeutic landscape, and a palettic approach, as a taken-for-granted way of understanding and categorising the environment in terms of its relationship to health. Likewise, in 2023, the UK government published its Environmental Improvement Plan which outlines "our vision for a quarter-of-a century of action to help the natural world regain and retain good health" (Department for Environment, Food and Rural Affairs [DEFRA], 2023), including the stated aim that "everyone should live within 15 minutes' walk of a green or blue space".

The palettic approach is also impacting health provision. In Scotland, "*Nature Prescriptions*" have been developed to provide "an innovative way to support the health of people and nature" through recognising the health impacts of exposure to natural space. The scheme was piloted in Shetland, and later in Edinburgh where five doctors surgeries across the city trialled the practice in November 2020. Reports from the pilot showed that of 350 patients given a nature prescription for a range of 32 conditions around three quarters reported that they had benefited (Greenspace Scotland, n.d.).

Pause for thought: *Consider how the critiques of therapeutic landscapes raised in this chapter might be applied to address the potential shortcomings or assumptions embedded in the examples discussed above.*

Acknowledgement of the complex relationship between health and landscape within policy discourse is significant and much academic and public work aims to quantitatively measure such health benefits (e.g. Gascon et al, 2017). It is important that scholars of the cultural landscape direct valuable critical attention towards such real-world applications of the therapeutic landscape concept and its allied approaches. Such criticality is crucial to interrogate the awkward assumptions and in-build exclusions which often characterise the implementation of such ideas.

Chapter Summary

Box 12.6 Key Concepts

Chapter 12

Socio-ecological models of health
The therapeutic landscape concept
Palettes of place
"Non-ordinary" landscapes
Critiques of therapeutic approaches

As Gesler noted early in his development of therapeutic landscapes research, "[a]ny solutions to the riddle of creating environments conducive to healing are necessarily complex" (1993: 173) (Box 12.6). This chapter outlines some of that complexity, beginning with Gesler's typology and charting the movement from research which focused specifically upon the non-ordinary, through a colour coding of "natural", wilderness and watery spaces, to a more critical consideration of "therapeutic" environments and experiences today. This field of study has shifted considerably over the last 30 years, moving to engage not only places which prove restorative but also those which actively promote wellbeing and sustain health. In doing so, scholars have highlighted the relational formation of therapeutic landscapes and emphasised that many benefits emerge from individual interactions between persons and places.

The therapeutic landscape is recognised as diverse and contingent, subjective and uneven. Therapeutic benefits emerge in relation to prior experience, temporal change or rhythms, and broader sensory stimuli and capacities. While definitions are inherently fraught entities, particularly given the development of this concept as outlined above, for the purposes of clarity in pulling together the main points of chapter, we offer the following from Milligan et al (2004: 1783):

> The concept of "therapeutic landscapes" is thus concerned with a holistic, socio-ecological model of health that focuses on those complex interactions that include the physical, mental, emotional, spiritual, societal and environmental.

Work on therapeutic landscapes combines and traverses many of the conceptual developments concerning cultural landscape study discussed throughout this textbook. Understanding those developments and trajectories helps appreciate the elaboration of this area of scholarship, as well as inform important scrutiny of the ways in which it is enrolled into policy and applied in practice.

Suggested Seminar Activities

Activity 1: Take three photos of a space you consider to be therapeutic. In groups, consider the following questions:

1. What makes these particular landscapes therapeutic to you?
2. Do they have distinctive features which correlate between them?
3. Do these features resonate with their peers/colleagues?
4. Do the therapeutic effects take time to appear?

Activity 2: Find and read a planning document (e.g. for a new town, community, school, hospital) and consider how the plan is infused with certain ideas of how landscapes can produce therapeutic effects. Consider:

1. Which aspects of the literature are drawn upon to discuss the production of therapeutic effects?
2. Which aspects are omitted?
3. Is the therapeutic experience seen as relational? Intersubjective?

Activity 3: Delve into the nuances of the therapeutic landscape literature through the following paper: Wilson K., (2003) "Therapeutic landscapes and First Nations peoples: an exploration in culture, health and place" *Health and Place* 9: 83–93. This paper enables discussion of the developments in the field of therapeutic landscapes in terms of the shift from "special" to "everyday" spaces as well as highlighting the Western focus of this body of work. Consider:

1. What is the difference between "special" and "everyday" space and why might this be important?
2. In what way does this paper relate to Conradson's ideas of a "therapeutic landscape experience"?
3. How does Wilson move beyond idea of physical space as therapeutic?
4. Why might "ways of knowing" have important policy implications?

Recommended Reading

Bell, S., Foley, R., Houghton, F., Maddrell, A., and Williams, A. 2018. From therapeutic landscapes to healthy spaces, places and practices: A scoping review. *Social Science and Medicine* 196: 123–130.

Foley, R., Kearns, R., Kistemann, T., and Wheeler, B. Eds. 2019. *Blue Space, Health and Wellbeing: Hydrophilia Unbounded.* London: Routledge.

Gesler, W. 1993. Therapeutic landscapes: Theory and a case study of Epidauros, Greece. *Environment and Planning D: Society and Space* 11: 171–189.

References

Arnold, D. 1993. *Colonizing the Body: State Medicine and Epidemic Disease in Nineteenth-Century India.* Berkley: University of California Press.

Atkinson, S. and Hunt, R. 2022. *GeoHumanities and Health.* Cham: Springer.

Bates, C. and Moles, K. 2024. Bobbing in the park: Wild swimming, conviviality and belonging. *Leisure Studies* 43: 887–899.

Bell, S. 2018. Sensing nature: Designing urban greenspace with sight impairment in mind. *Transactions of the Institute of British Geography* 44: 270–283.

Bell, S., Foley, R., Houghton, F., Maddrell, A., and Williams, A. 2018. From therapeutic landscapes to healthy spaces, places and practices: A scoping review. *Social Science and Medicine* 196: 123–130.

Biglin, J. 2020. Embodied and sensory experiences of therapeutic space: Refugee place-making within an urban allotment. *Health and Place* 62: 1–8.

Clark, G. and Dear, M. 1984. State apparatus and everyday life. In: Clark, G. and Dear, M. Eds. *State Apparatus in Everyday Life.* Boston: Allen and Unwin, pp. 60–82.

Conradson, D. 2005a. Freedom, space and perspective: Moving encounters with other ecologies. In: Davidson, J., Bondi, L., and Smith, M. Eds. *Emotional Geographies.* Aldershot: Ashgate Publishing Ltd., pp. 103–116.

Conradson, D. 2005b. Landscape, care and the relational self: Therapeutic encounters in rural England. *Health and Place* 11: 337–348.

Conradson, D. 2014. Health and wellbeing. In: Cloke, P., Crang, P., and Goodwin, M. Eds. *Introducing Human Geography.* 3rd ed. London: Routledge, pp. 599–612.

Conradson, D. 2016. Freedom, space and perspective: Moving encounters with other ecologies. In: Davidson, J., Bondi, L., and Smith, M. Eds. *Emotional Geographies.* Abingdon: Routledge, pp. 103–116.

Cosgrove, D. 1987. New directions in cultural geography. *Area* 19(2): 95–101.

Cronon, W. 1996. The Trouble with Wilderness: Or, Getting Back to the Wrong Nature. Environmental History 1: 7–28.

Davidson, J. and Parr, H. 2007 Anxious subjectivities and spaces of care: Therapeutic geographies of the UK National Phobics Society. In: Williams, A. Ed. *Therapeutic Landscapes.* London: Routledge, pp. 95–110.

Day, G. 2 January 2020. Why wild swimming is one of the best things to do for your mental health. *Vogue Australia.* Available at: www.vogue.com.au/beauty/wellbeing/why-wild-swimming-is-one-of-the-best-things-for-your-mental-health/image-gallery/50d7c3217ad7a0ad21ca6742e3ca8697

Department for Environment, Food and Rural Affairs (DEFRA). 2023. Environmental Improvement plan 2023: Executive summary. Available at: www.gov.uk/government/publications/environmental-improvement-plan/environmental-improvement-plan-2023-executive-summary

Duff, C. 2011. Networks, resources and agencies: On the character and production of enabling places. *Health and Place* 17(1): 149–156.

Dunkley CM. 2009. A therapeutic taskscape: theorizing place-making, discipline and care at a camp for troubled youth. Health Place. 15:88–96.

Ewert, A., Hollenhorst, S., McAvoy, L., and Russell, K. 2003. Therapeutic value of parks and protected areas. In: Harman, D. and Putney A. Eds. *The Full Value of Parks: From Economics to the Intangible*. Oxford: Rowman and Littlefield, pp. 141–156.

Foley, R. 2015. Swimming in Ireland: Immersions in therapeutic blue space. *Health and Place* 35: 218–225.

Foley, R. 2017. Swimming as an accretive practice in healthy blue space. *Emotion, Space and Society* 22: 43–51.

Foley, R. 2018. Palettes of place: Green/blue spaces and health. In: Crooks, V., Andrews, G., and Pearce, J. Eds. *Routledge Handbook of Health Geography.* London: Routledge, pp. 251–258.

Foley, R. and Kistemann, T. 2015. Blue space geographies: Enabling health in place. *Health and Place* 35: 157–165.

Foote, E. 2025. 'Plunging is the ultimate tonic': The invigorating power of cold-water swimming' *The Guardian*, 25 February 2025. Accessed online at www.theguardian.com/thefilter/2025/feb/21/invigorating-power-cold-water-swimming on 22/5/25.

Gascon, M., Zijlema. W., Vert, C., White, M., and Nieuwenhuijsen, M. 2017. Outdoor blue spaces, human health and well-being: A systematic review of quantitative studies, *International Journal of Hygiene and Environmental Health* 20: 1207–1221.

Gatrell, A. 2013. Therapeutic mobilities: walking and 'steps' to wellbeing and health. *Health and Place* 22: 98–106.

Gesler, W. 1992. Therapeutic landscapes: Medical issues in light of the new cultural geography. *Social Science & Medicine* 34: 735–746.

Gesler, W. 1993. Therapeutic landscapes: Theory and a case study of Epidauros, Greece. *Environment and Planning D* 11: 171–189.

Gesler, W. 1996. Lourdes: Healing in a place of pilgrimage. *Health Place* 2(2): 95–105.

Gesler, W. 1998. Bath's reputation as a healing place. In: Kearns, R. and Gesler, W. Eds. *Putting Health into Place.* Syracuse: Syracuse University Press, pp. 17–35.

Gesler, W., Bell, M., Curtis, S., Hubbard, P., and Francis, S. 2004. Therapy by design: Evaluating the UK hospital building system. *Health and Place* 10: 117–128.
Greenspace Scotland. n.d. Nature prescriptions: Supporting the health of people and nature. Accessed online at: www.greenspacescotland.org.uk/news/nature-prescriptions-supporting-the-health-of-people-and-nature on 22/5/25
Kearns, R. and Gesler, W. 1998. Introduction. In:. Kearns, R. and Gesler, W. Eds. *Putting Health into Place*. Syracuse: Syracuse University Press, pp. 1–13.
Kearns, R. and Moon, G. 2002. From medical to health geography: Novelty, place and theory after a decade of change. *Progress in Human Geography* 26(5): 605–625.
Lea, J. 2008. Retreating to nature: rethinking 'therapeutic landscapes'. Area. 40: 90–98.
Little, J. 2015 [2001]. Rural geography. In: Smelser, N. and Baltes, P. Eds. *International Encyclopedia of the Social and Behavioural Sciences*, vol. 20. Elsevier: Oxford, pp.795–800.
Milligan, C. and Bingley, A. 2007. Restorative places or scary spaces? The impact of woodland on the mental well-being of young adults. *Health and Place* 13(4): 799–811.
Milligan, C., Gatrell, A., and Bingley, A. 2004. Cultivating health: Therapeutic landscapes and older people in northern England. *Social Science and Medicine* 58(9): 1781–1793.
Morton, S. 2016. *Greenspace, Mental Wellbeing and Sustainable Communities*. UK Health Security Agency. Available at: https://publichealthmatters.blog.gov.uk/2016/11/09/greenspace-mental-wellbeing-and-sustainable-communities/
Parr, H. 2007. Mental health, nature work and social inclusion. *Environment and Planning D: Society and Space* 25: 537–561.
Philo, C. 2009. Medical geography. In: Gregory, D., Johnston, R., Pratt, G., Watts, M., and Whatmore, S., Eds. *The Dictionary of Human Geography*. 5th Ed. Chichester: Wiley-Blackwell, pp. 451–453.
Shortt, N., Rhynas, S., and Holloway, A. 2017. Place and recovery from alcohol dependence: A journey through photovoice. *Health and Place* 47: 147–155.
Strayed, C. 2012. *Wild: From Lost to Found on the Pacific Crest Trail*. London: Atlantic Books.
Thompson, M. 2021. The geographies of digital health: Digital therapeutic landscapes and mobilities. *Health and Place* 20: 1–8.
Thoreau, H. 1995 [1854]. *Walden or, Life in the Woods*. New York: Dover Publications.
Tuan, Y.-F. 1979. Thought and landscape: The eye and the mind's eye. In Meinig, D. W. Ed. *The Interpretation of Ordinary Landscapes*. New York: Oxford University Press, pp. 89–102.
Van Tulleken, C., Tipton, M., Massey, H., and Harper, C. 2018. Open water swimming as a treatment for major depressive disorder. *BMJ Case Reports* online first: 1–3.
Vannini, P. and Vannini, A. 2016 *Wilderness*. New York: Routledge.
Völker, S. and Kistemann, T. 2015. Developing the urban blue: Comparative health responses to blue and green urban open spaces in Germany. *Health & Place* 35: 196–205.
Wang. K., Cui., Q., and Xu, H. 2018. Desert as therapeutic space: Cultural interpretation of embodied experience in sand therapy in Xinjiang, China. *Health Place* 53: 173–181.
Ward-Thompson, E., Aspinall, P., and Bell, S. 2010. *Innovative Approaches to Researching Landscape and Health*. London: Routledge.
Whitworth, V. 2017. *Swimming with Seals*. London: Head of Zeus.
Wild. 2015. Directed by Jean-Marc Vallée. Pacific Standard.
Williams, A. 2007. Introduction: the continuing maturation of the Therapeutic Landscape concept. In: Williams, A. Ed. *Therapeutic Landscapes*. Hampshire: Ashgate Publishing Ltd., pp.1–14.

Wilton, R. and DeVerteuil, G. 2006. Spaces of sobriety/sites of power: Examining social model alcohol recovery programs as therapeutic landscapes. *Social Science & Medicine* 63(3): 649–661.

Winata, F. and McLafferty, S. 2023. Therapeutic landscapes and networks in restricted lives: Constructing restorative experiences among Indonesian female domestic workers in Hong Kong. *Wellbeing, Space and Society* 5: 1–11.

13 Into the Landscape

Introduction: Doing Landscape Research

The "cultural landscape" has many interwoven, complementary, or contradictory definitions, theorisations, and applications. It is neither fixed nor settled. Indeed, landscape, from the point of view of much scholarship, is a domain of *possibility*. Exploring such possibility requires attention to the doing of research. As Duncan and Duncan (2010) highlight, the questions we ask of a given landscape are often shaped by the way in we which define that term, the tools we use, and the subsequent information we gather through the process of research. Having showcased the various ways the cultural landscape can be encountered, interrogated, and understood, we turn now to how such environments are practically investigated. What are the tools that landscape researchers have at their disposal? Where might they deploy them? What are the issues to be considered in making the choice of what to use and where?

Chapter Outline

This is not a methods textbook, and our intention is not to offer an exhaustive account of all possible ways of doing landscape research. Rather, in closing the textbook, this chapter takes a thematic approach to discuss research methods in cultural landscape studies. Far from a comprehensive survey, each theme focuses upon a different way of doing research in conversation with the conceptual concerns outlined over previous chapters. These themes are "Talking About (and in) Landscapes", "Observing Landscape Encounters", "Archiving the Landscape / Landscape Archives", and "Representing and Responding to Landscape". Each section briefly contextualises relevant methodologies and their potential applications, before focusing down on one particular data collection method or research trajectory to give further insight into what each mode of enquiry might look like in the doing. Rather than provide seminar activities and readings at the end of the chapter, each section is followed by pedagogic prompts relevant to the approaches discussed.

DOI: 10.4324/9781003172451-14

Talking about (and in) Landscapes

Cultural landscapes concern human (as well as more-than-human) encounters with, and perceptions of, the environment. Consequently, we often need to *talk to people* to understand how cultures and surroundings interrelate. We might wish to talk to those who live in, or on, lands of interest to us; those passing through, visiting, or studying them; or even individuals distant from places, yet whose decisions ultimately shape the ways they appear to us. The value of informants for researching cultural landscapes has been long recognised. Carl Sauer often spoke with "local people"; individuals possessing vital insights or knowledge of value when trying to compose an account of a given area. As he writes, "[t]hey knew how to interpret the lay of the land, to keep a mental map, to note almost any change in the scene" (Sauer, 1956: 290).

Talking to people about landscape takes many forms. A formal (more or less) structured interview or focus group can involve participants specifically recruited because of knowledge or experience held, with questions or discussion themes carefully pre-crafted. Alternatively, more informal, everyday interactions might unfold as part of time spent getting to know a place. In each case, considerations of one's "sample" arise: who are we speaking to, why, and with what implications for the conclusions we might draw (or their wider applicability)? Much qualitative research (the kind that broadly concerns this textbook) is less concerned with representative accounts than *illustrative* ones (Valentine, 2013). The aim is often to capture diverse and varied experience to understand something of the multitude of perspectives or ways by which a cultural landscape works, is inhabited, or the processes and relationships which make it so.

There is also important question of where landscape conversations occur. The setting of a dialogue can, in significant ways, affect both conduct and content. Do we book a community space, approach informants at their offices, meet in a café or public place, or venture into the landscape to "walk and talk"? Indeed, the now-established qualitative method of the walking interview can take myriad forms (see Evans and Jones, 2011). At one extreme, the "go-along" interview serves as a "hybrid" of ethnography and conversation as researchers accompany subjects during normal or routine movements through an area (see Kusenbach, 2003). In contrast, researchers might lead or engage in guided walks (see Wang and Kao, 2017) and even produce scripted excursions to stage bespoke encounter with cultural landscapes for research (or artistic) purposes, for example via use of recorded audio prompts (Myers, 2010). Walking interview methods are heralded for "accessing a local community's connections to their surrounding environment" via elicitation of "place-specific" responses (Evans and Jones, 2011: 856–857). So, if we want to understand the role that former industrial mining landscapes play in terms of informing residents' sense of the rural (Wheeler, 2014); the significance of artistic depictions of distant environments in British Asian women's homes (Tolia-Kelly, 2016); or the practical ways in which the body responds to the rockface when climbing (Rickly, 2017), then attempting to do so in conversation amidst the environments in question can often prove effective.

Such approaches may also draw interviewees outwards, towards discussing their environment, rather than inwards, for deeper reflection or introspection (see Evans and Jones, 2011). The landscape can itself act as a "prompt" when it comes to recalling (or constructing) narratives of experience, memory, or history (Riley and Harvey, 2007: 395). As Michael Riley and David Harvey observe, drawing upon oral history interviews with nearly 90 British farmers in the Peak District and Devon, UK, as part of research exploring rural environmental change, being in the field was a vital generator of insight as "recollections were stimulated by, and grounded in, the landscape" (2007: 395). Specific topographical features or artefacts (an abandoned tractor, for example) served, for farmers, "as markers in the narrative" about broader historical processes or issues affecting their livelihood; as well as "a spur to several new trajectories within the life history" (2007: 396). Equally, a "multi-sensory experience of the land" was conducive to "intricacies", such as how the wind might affect the choice of cropping or management technique, to come to the fore in farmers' articulation of their knowledge of and relation to the landscape (2007: 396).

Suggested Activities

Activity: Watch the following video discussing the use of a "walking transect interview" in the context of a conversation between researcher Madelon Eelderink and Maria, a primary school teacher, working in an informal settlement in Kampala, Uganda.

- *Video Link:* www.youtube.com/watch?v=do7Nz1GOMqk&t=34s

Note the key points discussed about the walking transect approach as a method of participatory action research (PAR) and then produce

a. **a list of five to ten "top tips" or guidelines** for successfully utilising this participatory research method in the context of studying landscapes.
b. **as many "key considerations" as you can think of** (e.g. important questions or limitations), prompted by the video, that the researcher must reflect on when using/drawing from the data generated by this method.

Resource for further listening: The podcast "Talking Walking" presents walking interviews with artists, researchers, activists, and others who incorporate walking into their professional practice. Their conversations showcase the practice of the "walking interview" to great effect, as seen here: www.talkingwalking.net/

Observing Landscape Encounters

If cultural landscapes can be understood as, in part, expressions of those living in and acting upon them, then talking to people is not the only means of accessing information about cultural life. Methods of observation, both of landscapes and the encounters that take place within them, are another tool for research. Again, these practices have a long history; traceable to the centrality of distanced observation as key to scientific positivism and manifest in Carl Sauer's inductive methodology as it advocated the trained, "morphological eye" (see Chapter 1). Sauer's way of looking combined detachment from *and* physical immersion in the landscape. His commitment to fieldwork reflected his belief in the value of "becoming a participant observer of an unknown life and land" (Sauer, 1956: 296). Through field observations, understanding of landscape character (or "personality" – see Sauer, 1941) was constructed. While subsequent observational approaches to cultural landscape study have expanded interest in subjective experiences and participation, Sauer's empirical closeness clearly instilled the act of looking at the environment as a core research method. As Chapter 4 discussed, the study of "ordinary landscapes" as advocated by Donald Meinig (1979) (and typified in Pierce Lewis' [1979] axioms for "reading" cultural tastes and values from the environment as a "text") owes a clear debt to Sauer.

Nevertheless, as sociologist John Law (2004) argues, the world cannot be understood via looking alone. The ideal of the detached observer, prevalent in early landscape studies and enshrined in Western theories of knowledge since the Scientific Revolution, has been roundly critiqued not least *by* landscape scholarship (see Chapter 2)! A more immersive practice of participant observation, whereby involvement in cultural landscape activities enables attention to the mundane aspects of what we or others do, has emerged to embrace the opportunities afforded to the researcher by becoming engaged with the landscape (or phenomena) they wish to understand. Importantly, such research involves participating *and* observing. This necessitates being in "the field" to witness something – wherever that may be (a rural walking group in the hills, an urban protest camp, lurking in an online forum) – *and* actively engaging with the practices that shape it. As Eric Laurier (2016: 169) remarks, this is "perhaps the easiest method in the world to use because it is ubiquitous and we can all already do it", in that "[f]rom the moment we are born we are, in various ways, observing [...] and trying to participate in" the world. This kind of research can offer intimacy with a given subject (here, a landscape) and it is useful to emphasise both the active and grounded nature of such a data production process. Moreover, collaboration and involvement with those communities, groups, or places under scrutiny can constitute a more responsible and ethical scholarship (see Box 13.1).

Indeed, ethical considerations are key for observational practice, especially if covert means are employed. One must also consider the importance of reflecting upon positionality and subjectivity in terms of their shaping access to, and interpretations of, respondents and sites (McMorran and Green, 2023). How do we, as beings with bodily capacities or cultured dispositions, experience landscape in ways others may not? Being attentive to the partiality and particularity of carrying out research in particular places is paramount. For example, in her work on the laughter of walkers in the Lake District and Peak District of the UK, Macpherson (2016) emphasises how the endorphin-induced highs of walking outdoors might result in overly "positive dispositions" in both researcher and researched. As an able-bodied researcher working with partially sighted hikers, Macpherson is acutely aware of the limitations of her own perspective. One must be mindful of the awkwardness inherent to any generalising about landscape encounters, meanings, or practices based on our *own* experiences. In writing up "findings", much of our data may also take a form hard to express or translate into written text (see below). Remembering that "there is always a gap between lived experience and communication" (Hoggart et al, 2002: 210); both in terms of how respondents articulate relationships to landscape, and how we convey observations, prompts the seeking of alternative ways of representing cultural landscapes.

Box 13.1 Participatory Action Research and Landscape

Participatory action research (PAR) involves researchers working *with* respondents to develop project foci which responds to the needs, concerns, or interests of a particular community. Thus, such work aims to make a direct, tangible (positive) difference or contribution to participants' existence. This kind of research also challenges traditional hierarchies of researcher/research subject to create more collaborative approaches to enquiry. Often pursued in the context of studies of development or efforts to decolonise research praxis (yet it need not always be so), PAR raises questions about the kinds of outputs, motivations, dissemination activities, and priorities that researchers pursue. It is often avowedly – and unashamedly – political in its identification of clear issues, or "wrongs", that academic knowledge and expertise might be deployed (alongside other forms of knowledge and expertise) to address; reflecting and learning from the process to inform future endeavours.

Whilst an identified scholarly methodological approach since the 1940s, only in recent decades has PAR become a more mainstream practice. This reflects an increasing sense of the need for academic accountability, ethical research engagement, and the assertion of various efforts to critique the power structures that scholars are themselves culpable in propagating (Kindon et al, 2007). In the context of cultural landscape studies, PAR takes several forms, showcasing a variety of ways in which participation might happen. Importantly, PAR is not a "method" as such, but a framework for how research is conducted in collaboration with landscape residents, users, and communities. Thus, involving participants in the process of mapping landscapes through participatory planning workshops and discussions; storytelling approaches; autho-ethnographies, interviewee verification; producing creative responses or exhibitions (e.g. photography), all offer means, through landscape research, to address issues of relevance and empower communities in contexts of planning and landscape change. Such is the case with Amy Nettley and her collaborators' work on visualising sea-level rise in Cornwall, UK (Nettey et al, 2014). Advised and informed by residents and stakeholders at key stages of the research, the project drew from the community in order to inform how spatial models were applied and integrated into a public-facing film output.

Some PAR scholarship suggests scholars should go further. Sarah Radcliffe highlights that decolonial research extends understandings of participatory research "by getting participants input in research design as well as analysis and sharing of results" (2020: n.p.). Ensuring participation is intrinsic to the *entire* research process (from conception to dissemination) helps "de-center the researchers' individual autonomy", enabling research to "fulfil its decolonizing potential" (2020: n.p.). Moreover, centring participation throughout the entire research process not only generates new ideas and directions but tailors forms of *dissemination.* Radcliffe discusses work by the *Taller de Historia Oral Andina* (THOA, Andean Oral History Workshop) where participants were involved in the generation of outputs, leading to the production of "an Aymara-language radio soap opera, as that was the main media source for rural low-income groups" (2020: n.p.). Such a format for conveying their findings may never have occurred to the researchers if they had not involved a wider set of individuals beyond the academy.

Working within the considerations outlined above, participant observation is a key method for cultural landscape study from a range of standpoints. However, such learning through looking and doing is perhaps most often used in work variously labelled non- or more-than-representational. This is because, as a method, participant observation is particularly well-suited when seeking to understand events, relations, practices, performances, atmospheres, or place feelings which "fall outside of common awareness" (Vannini, 2015: 9). Thus, research via this method has explored the ways in which the world is variously animated or mediated by nonhuman and non-cognitive processes (e.g. outside of the world of human categorical thought), as well as how such nonhuman or non-representational forces are often *managed* to create a particular felt, lived experience of landscape. Sometimes this involves research done with other creatures (see Box 13.2).

Box 13.2 Multispecies Ethnography and Landscape Study

Multispecies ethnography has been defined as "ethnographic research and writing that is attuned to life's emergence within a shifting assemblage of agentive beings" (Ogden et al, 2013: 6). Such research involves, often, participatory observation specifically concerned with "the relations of people, animals, plants, and things" (Ogden et al, 2013: 5) as they define or help explain a particular context or landscape. Where underpinned by relational or posthuman approaches to nature and culture (see Chapter 10) such works consider nonhumans as "agents" in social and environmental processes, whose activities, and interests, matter in what happens. Crucially, humans (our experiences, practices, spaces, and politics) are figured as being "more-than-human", always (re)made through relationships with all manner of other entities too often bracketed out of analysis.

In terms of researching landscapes, such research often focuses on the ways in which humans and nonhumans engage with each other in particular "contact zones" (after Haraway, 2008), changing each other as a result. For example, Adam Keul's (2013, see Chapter 10) work on alligators in Louisiana swamplands attends to the ways in which interactions between humans (as tourists or boaters) and these reptiles "produce" versions of the landscape differently dangerous, exciting, or economically productive.

One example of where participatory practice has recently flourished in landscape studies is in explorations of how environments are sensed. For instance, environmental histories of sound and noise (Coates, 2005) or the lived sensitivity to heat in the context of off-grid living (Vaninni and Taggart, 2014) have foregrounded the body's sensory array in other ways. In some projects, the sensory landscape encounter has been used to explore smells as "olfactory histories" which influence landscape transformation through time (Parsons and Fisher, 2021: 29) and produce distinct "smell-scapes". Similarly, Allen (2021) uses the smell walk as method to understand the smellscape as the "olfactory equivalent of a landscape" (Allen, 2021: 23). First coined by Porteous (1985), researching a smellscape requires the researcher to engage in "smell catching" (McLean, 2019), whereby the researcher takes pause at a new smell, breathes deeply, and seeks to put smell at the forefront of their senses. These experiences are simultaneously captured on a mapping app.

Therefore, while the outcome of research that foregrounds the senses affords a different way of knowing cultural landscapes, the practical "doing" has much in common with other observational techniques. For example, in his work on therapeutic landscapes, Gorman uses smartphone "scratch notes" to quickly record in-situ his "visual, tactile, aural, olfactory observations of the farms, as well as those of… [his] participants" (2017: 24). The task of the researcher, across diverse forms of participant observation, concerns noticing and recording (electronic, verbal, or written). Indeed, this act of translation (from encounter to record) marks the first stage of analysis as the researcher selects those elements of their experience being of scholarly interest. In this way, privileging smell, sound, touch, or sight; like selecting certain sections of text from a document, or words in an interview, is an active choice, revealing as much about the researcher (their biases, knowledge, capacities, personal interests) as their study's focus.

Suggested Activities

Activity 1: Read the following: www.ericlaurier.co.uk/resources/Writings/Laurier_Part_Ob_V2.pdf?

- Write a short definition of participant observation, if in a group then discuss your interpretations. Do they match?
- Individually consider (or discuss in a group) the difference between "outsider" and "insider" as discussed by Laurier, and the implication of these categories for how such a method might work.
- Compile a list of the key strengths and limitations of this method, supplementing this with as many "key considerations" or "top tips" that might make for a more successful project using participant observation.

Activity 2: Go to a landscape and compile a series of observations on one of the following themes:

- *Landscape and identity;*
- *Power and regulation;*
- *Bodies and sensory knowledge.*

For your chosen theme, devise a set of clear criteria or prompts to help direct or organise your observations, as well as a structured means of documenting what you observe (a table or recording sheet with headings). Summarise your findings and note how they could illustrate key ideas within cultural landscape study.

> ***Suggestion:*** *Undertake this activity twice – once in a "familiar" landscape and again in an "unfamiliar" landscape, and have students reflect on the different experiences.*

Archiving the Landscape/Landscape Archives

For geographer David Matless, "all studies of landscape entail historical geography" (2003: 230). After all, the landscape appears as a *palimpsest* of human (and non-human) processes and activities, overlain throughout history. Despite landscapes being built up from various "traces", different features of these are attended to, highlighted, or celebrated to serve political interests. As Barbara Bender acknowledges, "[s]ome pasts are more audible than others" (2001: 4). A historical perspective on landscape offers important means for recovering "that which has been hidden from view" (Morin, 2009: 289), including the lives or experiences of those who (perhaps under duress) made scenes what they are today.

Aside from engagements with those who live(d) in or around our cultural landscapes of interest, who may hold memories and information (e.g. oral histories) shedding light on how environments came to be, the scholar of landscape seeking to know its formation through time is likely drawn to the *archive*. This term often refers to a collection of materials of some kind, unified by association with a particular set of historical events, processes, individuals, or periods. An archive might be publicly accessible, institutionally managed, and highly categorised. Alternatively, it might consist of documents made, held, and arranged by individuals on an informal, personal basis, a box in a basement or sheaf of papers in a loft.

Importantly, they are not only made up of *written* documents, though these may form the bulk of what researchers encounter. Images, sound recordings, artefacts, and objects (like old instruments or natural history specimens) are some examples of things contained in some archives. Thus, working within archival collections, both conventional and not, reveals "the archive" not to be any settled *thing*, but instead a provisional, unfinished, contingent, amorphous gathering of "stuff" brought together by the very act of research (see Lorimer, 2010).

When reflecting upon archival practice, Jacques Derrida offers some useful observations on the archive as both a physical site that gathers materials from myriad other places and times; *and* an active mechanism of power, since this gathering can serve important institutional function in the present (and future) (Derrida, 1995). Consequently, organised archives both reveal social relationships and express institutional authority over those relationships. What is kept, considered worthy, where, and by whom are key questions of power associated with an archive. Thus, the archive is, as Dwyer and Davies (2010: 89) propose, a site of "contradictory processes", which re-animates the lives, places, and experiences of the past whilst simultaneously masking those deemed uninteresting, dangerous, or not worth remembering.

These questions, about which documents, stories, accounts, experiences, or forms of knowledge are included or excluded, have consequences for the cultural landscape histories told (or what we even consider to *be* such histories) (see Chapters 6, 7, and 11). They also prompt debate over how archival collections should be managed and housed, whether they might be made (more) accessible to the communities whose stories they contain, and how collaborative forms of archive management might lead to emancipatory partnerships and ways of acting in *contemporary* landscape surroundings (see Tindley et al, 2019). As Bender writes, "[w]e make the past in the present. We set the agendas and often we use the past to justify the present. [...] there isn't one history, but multiple histories – and some are legitimised and others excluded" (1998: 18–22).

Box 13.3 Enlivening a Militarised Landscape

Photographer Alex Boyd and archaeologist Lesley McFadyen (see Boyd and McFadyen, 2024) demonstrate via recent work how the landscape, *as an archive*, can be creatively engaged to recover lived histories otherwise unrecorded within institutional documents or repositories. Their work draws on over 20 years of Boyd's photographic practice within the landscape of the Ardeer Peninsula, North Ayrshire, UK. On the peninsula there was (and, to an extent, remains) a munitions factory, established in 1871 by Alfred Nobel to manufacture dynamite. It was later developed by Imperial Chemical Industries to produce other explosives, including cordite in the 1940s before the factory site was shuttered in the 1990s. Today, the site is earmarked for brownfield development; though local and national environmental campaigners have argued that its sand dune ecology, partly created by the engineering of blast defences, deserves conservation and protection. For Boyd and McFadyen, there is more than ecological value in play here. They seek to recover something of the landscape's lost labour histories, particularly concerning its majority female workforce of "munitionettes" who staffed the factory during two world wars during the twentieth century.

Boyd documents numerous inscriptions by workers onto the *fabric* of the environment itself, demonstrating how landscape photography can help recover erstwhile ignored archaeological and lived histories, unearthing alternative heritages and narratives. Whilst the shape of the site speaks to a history of environmental engineering in service of industrial military activity, it has also been overlain with "the material presence of others" (Boyd and McFadyen, 2024: 11) in the form of pencil graffiti (some dated to 1941) comprising portraits, lists of names, and song lyrics, inscribed by the munitionettes. Boyd and McFadyen focus on "fifteen songs, written out by hand" (2024: 12) on painted and plastered walls, many still visible, to argue that, more than evidence of lyrics, the graffiti "is evidence for singing" (2024: 13). Thus, these landscape traces read as archaeological evidence enable working with landscape-as-archive to recover *embodied* forms of living (and working). Making the munitionettes a presence in this "aural-scape" challenges their absence (and silence) within more conventional, institutionally held archival records. As Boyd and McFadyen reflect upon their practice:

> this is not a story or landscape that ends with decay, and an archaeology of photography creates a new kind of accounting for memory and place without passing over working landscapes, whilst also recognising the importance and force of other-than-human lives and the future direction and possibilities of a more than human landscape.
>
> (2024: 15)

Crucially, a landscape (and the deposits or material remains which make it) can also be considered *an archive* (see Box 13.3). For Hayden Lorimer, direct encounters with landscapes often provoke thinking about their historicity:

> Hip-high, shoulder deep or skein slim, past landscapes are about coming to the spot. Details of location can be cross referenced. Things can be scooped up, delved into, even dug up. A less settled phenomenon than the historian's palimpsest (of clean-cut layers resting level upon level) and enlivened by presence, landscapes promise depth and disruption.
>
> (2010: 257)

Certainly, we should be cautious of claiming to know for certain, or of extrapolating, based solely on what, for Cole Harris, is often "a thin and jumbled record that can be exceedingly difficult to read" (2001: 330). Nevertheless, field observations complement, provide ways into, and help make tangible the content of various archival resources, raising new questions and triangulating (or contesting) our interpretations. Archives help us go beyond simply "looking at the land" (Harris, 2001), but that does *not* mean such acts of looking are not worthwhile. Indeed, by looking at the environment *and* perusing documents, photos, and other materials, we can build an appreciation of landscape "as a slowly built up as well as an actively made built form"; in turn uncovering vital stories about who, through the course of lives lived, made it what it is (Mitchell et al, 2021: 27).

Working in archives takes time. Engaging these (sometimes vast) stores of information demands "patient labours" (Lorimer, 2010: 251). Archives are not to be *dabbled* in but necessitate (and reward) more substantial commitment. Encounters with archives can be bewildering or baffling. The first time one sits and is confronted by a mass of textual (and non-textual) records, even those most carefully catalogued and organised, can be terrifying for even the most seasoned researcher. But, with time, effort, and reflection, patterns often appear as threads are traced and followed; stories pieced together. Would-be researchers of cultural landscapes should not be put off by the wide array of information available, nor fall into the trap of thinking that one, single, correct historical narrative awaits within dusty, yellowed pages. Rather, it becomes the job of the scholar to make sense of this jumble in conversation with relevant concepts, frameworks, and research questions and to work with the archive to tell its story. As Matless reminds us, "[a] different history implies a different landscape, and a different way forward" (2003: 230).

Suggested Activities

Activity 1: Formal archives are likely unfamiliar to most students and it is useful to visit and familiarise yourselves with a specific archive, either online or in person. Identify one such archive, whether at national/state level or housed within a local or university library. Consider working through the stages that lead from identifying an archive's location, entering the collection, and finally encountering a physical or electronic "document" before you. This may include exploring archive catalogues, negotiating levels of access, or discussing the arrangements for reproducing and copying material. If discussion is occurring within a group setting, consider printing off a selection of documents and a section of archival catalogue to act out this exercise within the bounds of the classroom.

Activity 2: The following article explores the visual representation of the Uluru-Kata Tjuta and Tongariro National Parks in Australia and New Zealand following the transformation of these countries into postcolonial nations in the late twentieth century.

- Walliss J. 2014. Transformative landscapes: Postcolonial representations of Uluru-Kata Tjuta and Tongariro National Parks. *Space & Culture* 17(3): 280–296.

Read this article with a focus on the materials examined. Consider the sources of the images included and how they are used in analysis and to support the argument made. What kinds of archives has the author drawn material from? What does the paper reveal about the value of *non*-textual sources when doing so? How does this example speak to the value of working both within and *between* different kinds of collections (or archives), such as formal library resources, websites, and personal document stores?

Activity 3: Create your own alternative archive for a landscape of your choice. Gather the resources to archive it – you can create these or imagine them if required (personal diaries, postcards, policy documents, posters, newspaper articles, books, letters, social media posts visual and text based, podcasts, film). Next consider whose voices are being included in this archive. What perspectives are being held? What might a landscape researcher in 100 years' time be able to do with this collected material?

Representing and Responding to Landscape

As previous chapters considered, cultural landscape studies are often interested in interrogating examples of how the world is represented. Such representations take different forms. The imagery of romantic landscape painters can be unpicked in terms of the aesthetic tropes they (re)instate through their circulation and consumption, with implications for how today's "natural" landscapes are valued or conserved (e.g. Ellison, 2013). Equally, the depiction of landscapes, their features, history, and normative character via literary works or popular travel writing informs attention to the textualisation, discursive formation, and colonisation of diverse environments. Our imaginative and moral geographies of landscape owe much to written and visual culture, nowhere more obvious than the case of the "Orient" (Said, 1995).

In terms of *how* such scholarship uses representational materials, one can draw from the wealth of practices grouped under the heading "visual methodologies"; particularly when research engages paintings, photographs, maps, or other graphic representations of cultural landscapes. Gillian Rose (2022) has influentially defined the parameters of a "critical visual methodology", central to which is a twin concern with the many "sites" and "modalities" affecting the production and consumption of visual content. In terms of sites, these include the situation of the image's production (where, by whom, or in what context it is created); the image itself as it is encountered and responded to (e.g. its surface, what it depicts, how it appears); the places through which images circulate; and where images and various "audiences" or users come to interact. These sites are mediated by "modalities" including *technology* (in both making and distributing or [re]presenting); *composition* (in terms of how the content of images are organised, or images are arranged in relation to each other); and *social relations* (including the political, economic, cultural, and even environmental relationships that contextualise how we encounter and respond to visual materials).

Box 13.4 Unlikely Hikers in the Landscape

Tourism and leisure scholar Phiona Stanley's examination of "a nascent, primarily online community of so-called 'unlikely' hikers" based in the US explores how active involvement in outdoor recreation "contests mainstream representations in outdoors media, advertising, and social imaginaries" (Stanley, 2020: 242) and challenges hiking's "unbearable whiteness" (244). This community, comprising solo female and lesbian hikers, "fat hikers", and "hikers of colour", involves those often marginalised (or unlikely to feature) in the discursive and representational construction of the US National Park landscape. As recreational spaces, such sites are presented as being for a particular "outdoorspeople": "White, male, straight, muscular, and able bodied" (Stanley, 2020: 244).

Stanley's "netnographic" analysis of publicly accessible *Instagram* posts, produced and shared by "unlikely hikers", evidences the normative depictions of hiking culture; the acts of exclusion endured by those out with such norms; and the efforts of some in this community to foreground, celebrate, and advocate the presence of difference "on trail". Thus, hikers' online activities offer counter-representations through their efforts to "amplify each other's voices as they go out of their way to include one another" (Stanley, 2020, 251), "nudging at social imaginaries" by "making the 'rules' of hiking trails visible [and...] pushing back" (Stanley, 2020: 253). Stanley's work therefore emphasises the value of attending to contemporary social media through which such representations circulate, are encountered, and become reworked.

Different analytical approaches abound, including historical accounts of image production; semiotic, discursive, or compositional analyses; and examining or eliciting the production of images from informants to understand their worlds (see Pink, 2021 on various forms of "visual ethnography"). Each illuminates different dimensions of how sites and modalities mediate one another. Importantly, visual materials are "never innocent", requiring "critical" attention to "the agency of the image, [...] the social practices and effects of its circulation and viewing, and [...] the specificity of that viewing" for different ends (including academic research) (Rose, 2022: 45). A wide variety of representational materials exist through which we depict, encounter, or come to know our surroundings in different ways, and the rise of digital technologies and social media platforms opens additional formats, sites, and relations for consideration. This is particularly well illustrated through recent work that engages with representations of landscape via online communities (see Box 13.4).

It is also worth appreciating how creative ways of writing, depicting, or otherwise representing landscape have been deployed as a means of doing/disseminating research. For example, geographers, as part of a burgeoning disciplinary "orientation towards embodied and practice-based doings" (Hawkins, 2015: 248), have used sketching, drawing, and painting as a means of staging a research encounter with place. Drawing practices have also been embraced and experimented with in other areas of the humanities, such as anthropology (see Ingold, 2011). Working in collaboration with artists and/or participants can lead into new ways of articulating landscapes' meaning and imaginary manifestation, as well as forms of public presentation, such as through exhibitions (e.g. Tolia-Kelly, 2016). In reconfiguring drawing as a "phenomenological" act, such representational practices become a means to access and reflect on landscape's *more-than*-representational constitution (Wylie and Webster, 2019).

In a similar vein, creative strategies of *writing*, informed by literary approaches, have also been used to document the fleeting, felt, lived character of landscapes, for example via engagements with a resurgent nature- or place-writing tradition (see Lorimer, 2019). The work of anthropologist Kathleen Stewart is of note here: her "ficto-critical" works utilise vignettes and experimental forms (lists, tangential meditations, stories, written sketches) to, as she puts it, "sidle up to [a region's] peculiar ongoing generativity" in ways that bring "the material and the ephemeral together" (2013: 283–284). Finally, poetry, too, has become a fertile area for expression taken forward by some in discussions of landscape. Notable here is the work of Hywel Griffiths (e.g. Griffiths et al, 2017), whose analysis of medieval Welsh "praise poetry" reveals such materials as valuable for informing insights regarding historic community relations and responses to flood events and management. Thus, creative acts of *writing*, as well as visualising, cultural landscapes offer fertile ground for research into how we see and experience our surroundings.

Suggested Activities

Activity 1: Seek out an example of regional or national tourist media (e.g. a leaflet, advert, information sign, or website). Consider the images of cultural landscape in your chosen example: how do they (re)produce particular narratives about that landscape, its history, its people, its ecology? For example, consider the depictions of destinations hosted on the *Visit Brasil* webpages: www.visitbrasil.com/en/

Activity 2: In Chapter 2 (Box 2.4), we introduced the work of artist Kehinde Wiley, which has served to recontextualise the aesthetics of landscape. Consider the landscape images included in this review of his National Gallery exhibition: www.athenaartfoundation.org/kehinde-wiley – particularly those that reinvent or reinterpret the work of paintings by artist Casper David Friedrich. What is the *effect* of what Wiley is doing with these images? What does Wiley's work reveal about the politics of landscape representations, and who the landscape is for?

Into the Field

In this final section, we reflect on and share some of our experiences when it comes to undertaking (and teaching) landscape research in the field. These points are intended to support students *and* lecturers/practitioners working on topics of landscape.

Doing research is a practical matter and, as such, raises practical questions. Questions are as follows: what regulations (institutional or legal) do I need to abide by, or work within (and with what effects)? How might access be negotiated to engage the places, people, or stores of knowledge that I need to understand my topic? Are there particular forms, processes, or conversations that need to be had to secure such access; permissions regarding what information might be *taken* back to our desks; or agree the mechanisms for consent, data protection, and dissemination that both respect and protect the rights and interests of respondents? In acknowledgement of the need to reflect on and challenge the normative ideas of fieldwork (and the fieldworker) that have long characterised disciplines such as anthropology or geography, we should also question what barriers our choices (perhaps inadvertently) erect with regard to *who* can participate in the research process, and on what grounds. The capacities of one's body; the kinds of epistemologies (ways of knowing) or ontological foundations (basic assumptions about the world as it is) that individuals bring to bear; the ways specified differences affect or rework encounters with sites and social groups: all demand consideration if we are to incorporate as varied a set of perspectives and (student or researcher) experiences as justifies the diversity of cultural landscapes. In other words, we need to be mindful of the many overt and implicit ways in which exclusion can manifest in the field.

When it comes to organising field excursions or site visits, the co-design of student trips *with* students (like the co-design of research projects with interested stakeholders – see Box 13.1) can result in greater class buy-in and learning experiences which meet the needs and interests of all involved. There is, more broadly, an encouraging increase in relational pedagogy which recognises the need to reimagine the more transactional approach to learning in contemporary higher education (Biesta, 2006) and we argue that this co-creation can, and should, happen outside of the classroom. Developing fieldtrips with students could incorporate suggestions or workshops to flesh out ideas such as what it's for; where it might take place; how it could be experienced; and, if appropriate, the ways in which it connects to assessment. This relational approach might even be future facing (design a fieldtrip for the next cohort) or speculative (imagine your ideal fieldtrip). Ultimately, engaging students into the design of field research draws them into the process of their own learning and opens conversations about how research happens. Of course, we recognise the practicalities at play too. Teaching requires planning and bureaucracy is often accompanied by deadlines. Relational teaching is not easy, but we encourage the embracing of its pedagogical potential.

When it comes to the encounter itself (with a designated environment, social group, archive or site of interest), it is important to think about how different theoretical understandings of "landscape" might "orient" (after Ahmed, 2006) us to its different features, inhabitants, characteristics, or affects. The (scholarly) language used to frame a trip, articulate its aims and structure, or set questions and activities; alongside requests to read this or that article, or complete a given task prior to an outing, all shape how a destination is made sense of. It is imperative we provide as many *openings* onto a given landscape as possible; that we avoid foreclosing on its possibilities to enliven, connect with, or speak back to different ideas and debates. In our own work, we have modelled such an engagement with the landscape via a visit to the Yorkshire Sculpture Park, UK (Garlick and Hunt, 2022), and this approach might be trialed, or adapted, within other locations.

There are also considerations of what comes after the field. What do teachers or students do with the products of that experience? Several of the activities we have elaborated during this chapter explicitly stage a direct encounter with landscape. Others could easily be transposed beyond the classroom or enlivened by incorporating material gathered via site visits or data collection. In addition to any assessed or formative work aiming to apply key ideas or methods in situ, incorporating time for reflection after fieldwork is also valuable. This may take the form of written submissions, creative artefacts, or presentations intended as either formative or summative. Structured prompts can encourage connections with key learning outcomes, or a recognition of transferable and disciplinary skills and knowledges acquired (or applied). Reflections might also incorporate introspection: how did one's experiences in relation to an environment, context, social group, particular issue, or observed phenomena make students *feel?* How does such feeling infuse, animate, motivate (or inhibit) understanding, engagement, or their wider educational experience?

Ultimately, as Anna Guasco (2022) articulates, rather than justifying only decisions *not* to do fieldwork, it is advisable to consider seriously our reasons for "going there", especially when such activity "is neither carbon neutral nor socially neutral" (468–469). As she concludes, we must rethink our disciplinary tendency to reify field-working as the only or best way to do research (geographical or otherwise) and, we suggest, education:

> An "ethic of not going there" does not mean that fieldwork and travel would never be okay, but rather that they would never be the presumed baseline or norm, the supposedly neutral standard option. It means that doing fieldwork must be justified in the same way that not doing fieldwork is expected to be justified.
>
> (Guasco, 2022: 472)

Chapter Summary

Box 13.5 Key Concepts

Chapter 13

Talking methods (interviews, focus groups, etc.)
Participant observation/ethnography
Archives
Visual methods
Landscape fieldwork

This chapter has outlined many ways in which cultural landscapes have been practically researched (Box 13.5). This has included the various ways in which researchers can talk about (or in) landscapes through the interview, focus group, or walking interview. Turning to observations, we detailed how landscape can be looked upon, participated in, and sensed. In addressing archival methods, we reflected on the various ways in which diverse histories of landscape can be uncovered and reconstructed through engagement with archives of all kinds: formal and informal, institutional and personal, textual and material, repositories and libraries, or even the landscape itself. Penultimately, different strategies for representing and responding to landscapes (and how such representations are critically analysed) were explored, with emphasis on burgeoning creative and non-traditional forms of representation (including creative writing, artistic practice, and social media). Finally, we discussed the doing of landscape research in the field, acknowledging some of the practical, bureaucratic, and pedagogic concerns at play.

Our discussion of landscape methods is by no means exhaustive. We encourage you to seek out further examples of methods in use. Scholars continue to find novel, creative means to engage cultural landscapes, and recombine or rework existing methodological approaches in new contexts. It is for the student of landscape to decide which of these suits their research purposes. As Duncan and Duncan remind us, "[d]iffering ways of answering the question 'What are landscapes?' will lead to differing ways of studying them" (2010: 226).

References

Ahmed, S. 2006. *Queer Phenomenology: Objects, Orientations, Others*. Durham, NC: Duke University Press.
Allen, L. 2021. The smell of lockdown: Smellwalks as sensuous methodology. *Qualitative Research* 23(1): 21–37.
Bender, B. 1998. *Stonehenge: Making Space*. London: Bloomsbury.

Bender, B. 2001. Introduction. In: Bender, B. and Winer, M. Eds. *Contested Landscapes*. Berg: Oxford, pp. 1–18.

Biesta, G. 2006. *Beyond Learning: Democratic Education for a Human Future*. Boulder: Paradigm Publishers.

Boyd, A. and McFadyen, L. 2024. 'The melody that's sweetly played in tune' – The heritage ecologies of the Ardeer Peninsula, North Ayrshire, Scotland. *Landscape Research*. 50:368–385.

Coates, P. 2005. The strange stillness of the past: Towards an environmental history of sound and noise. *Environmental History* 10(4): 636–665.

Derrida, D. 1995. Archive fever: A Freudian impression. *Diacritics* 25(2): 9–63.

Duncan, J. and Duncan, N. 2010. Doing landscape interpretation. In: DeLyser, D., Herbert, S., Aitken, S., Crang, M., and McDowell, L. Eds. *The SAGE Handbook of Qualitative Geography*. London: Sage, pp. 225–247.

Dwyer, C. and Davies, G. 2010. Qualitative methods III: Animating archives, artful interventions and online environments. *Progress in Human Geography* 34(1): 88–97.

Ellison, A.M. 2013. The suffocating embrace of landscape and the picturesque conditioning of ecology. *Landscape Journal* 32(1): 79–94.

Evans, J. and Jones, P. 2011. The walking interview: Methodology, mobility and place. *Applied Geography* 31: 849–858.

Garlick, B. and Hunt, R. 2022. A visit to the Yorkshire sculpture park. In: Ballamingie, P. and Szanto, D. Eds. *Showing Theory to Know Theory: Understanding Social Science Concepts through Illustrative Vignettes*. Showing Theory Press. https://doi.org/10.22215/stkt/gb62

Gorman, R. 2017. Smelling therapeutic landscapes: Embodied encounters with spaces of care farming. *Health and Place* 47: 22–28.

Griffiths, H., Salisbury, E., and Tooth, S. 2017. "May God place a bridge over the River Tywi" Interrogating flood perceptions and memories in Welsh medieval poetry. In: Endfield, G. and Veale, L. Eds. *Cultural Histories, Memories and Extreme Weather a Historical Geography Perspective*. London: Routledge, pp. 93–111.

Guasco, A. 2022. On an ethic of not going there. *The Geographical Journal* 188(3): 468–475.

Haraway, D. 2008. *When Species Meet*. Minneapolis, MN: University of Minnesota Press.

Harris, C. 2001. Archival fieldwork. *Geographical Review* 91: 328–334.

Hawkins, H. 2015. Creative geographic methods: Knowing, representing, intervening. On composing place and page. *Cultural Geographies* 22(2): 247–268.

Hoggart, K., Lees, L. and David, A., 2002. Researching Human Geography. London: Arnold.

Ingold, T., 2011. 'Introduction' in Redrawing Anthropology: Materials, Movements, Lines. Ed. Ingold, T., Routledge: London. pp. 1–20.

Keul, A. 2013. Performing the swamp, producing the wetland: Social spatialization in the Atchafalaya Basin. *Geoforum* 45: 315–324.

Kindon, S., Pain, R., and Kesby, M. 2007. Participatory action research: Origins, approaches and methods. In: Kindon, S., Pain, R., and Kesby, M. Eds. *Participatory Action Research Approaches and Methods: Connecting People, Participation and Place*. New York: Routledge, pp. 9–18.

Kusenbach, M. 2003. Street phenomenology: The go-along as ethnographic research tool. *Ethnography* 4(3): 455–485.

Laurier, E. 2016. Participant and non-participant observation. In: Clifford, N., Cope, M., French, S., and Gillespie, T. Eds. *Key Methods in Geography*. 3rd Ed. London, Sage, pp. 169–181.

Law, J. 2004. *After Method: Mess in Social Science Research*. London: Routledge.

Lewis, P. 1979. Axioms for reading the landscape: Some guides to the American Scene. In: Meinig, D. Ed. *The Interpretation of Ordinary Landscapes*. Oxford: OUP, pp. 11–32.
Lorimer, H. 2010. Caught in the nick of time: Archives and fieldwork. In: DeLyser, D., Herbert, S., Aitken, S., Crang, M., and McDowell, L. Eds. *The SAGE Handbook of Qualitative Geography*. London: Sage, pp. 248–274.
Lorimer, H. 2019. Dear departed: Writing the lifeworlds of place. *Transactions of the Institute of British Geographers* 44(2): 331–345.
Macpherson, H. 2016. Walking methods in landscape research: Moving bodies, spaces of disclosure and rapport. *Landscape Research* 41(4): 425–432.
Matless, D. 2003. Introduction In: Anderson, K., Domosh, M., Mile, S., and Thrift, N. Eds. *Handbook of Cultural Geography*. London: Sage, pp. 227–232.
McLean, K. 2019 Nose First: Practices of Smellwalking and Smellscape Mapping. PhD Thesis, Royal College of Art, London, UK.
McMorran, C. and Green, N. 2023. Participant observation. In: Clifford, N., Cope, M., Gillespie, T., and French, S. Eds. *Key Methods in Geography*. 4th Ed. London, Sage, pp. 197–212.
Meinig, D. Ed. 1979. *The Interpretation of Ordinary Landscapes*. Oxford: OUP.
Mitchell, D., Jönsson, E., and Pries, J. 2021. Making the people's landscape: Landscape ideals, collective labour, and the people's parks (Folkets Parker) movement in Sweden, 1891–present. *Journal of Historical Geography* 72: 23–39.
Morin, K. 2009. Landscape: Representing and interpreting the world. In: Clifford, N., Holloway, S., Rice, S., and Valentine, G. Eds. *Key Concepts in Geography*. 2nd Ed. London: Sage, pp. 286–299.
Myers, M. 2010. 'Walk with me, talk with me': The art of conversive wayfinding. *Visual Studies* 25(1): 59–68.
Nettley, A., Desilvey, C., Anderson, K., Wetherelt, A., & Caseldine, C. 2013. Visualising Sea-Level Rise at a Coastal Heritage Site: Participatory Process and Creative Communication. Landscape Research, 39: 647–667.
Nettley, A., DeSilvey, C., Anderson, K., Wetherelt, A., and Caseldine, C. 2014. Visualising sea-level rise at a coastal heritage site: Participatory process and creative communication. *Landscape Research* 39: 647–667.
Ogden, L., Hall, B., and Tanita, K. 2013. Animals, plants, people, and things: A review of multispecies ethnography. *Environment and Society: Advances in Research* 4(2013): 5–24.
Parsons, M. and Fisher, K. 2021. Historic smellscapes in Aotearoa New Zealand: Intersections between colonial knowledges of smell, race and wetlands. *Journal of Historical Geography* 74: 28–43.
Pink, S. 2021. *Doing Visual Ethnography*. 4th Ed. London: Sage.
Porteous, D. 1985. Smellscape. *Progress in Physical Geography* 9(3): 356–378.
Radcliffe, S. 2020. Decolonising remote research. *Blog*. Available at: www.geog.cam.ac.uk/research/readinggroups/fieldwork/decolonise/ accessed on 27 August 2025.
Rickly, J.M. 2017. The (re) production of climbing space: Bodies, gestures, texts. *Cultural Geographies* 24(1): 69–88.
Riley, M. and Harvey, D. 2007. Oral histories, farm practice and uncovering meaning in the countryside. *Social and Cultural Geography* 8(3): 391–341.
Rose, G. 2022. *Visual Methodologies*. 5th Ed. London: Sage.
Said, E. 1995. *Orientalism*. London: Penguin.
Sauer, C.O. 1941. The personality of Mexico. *Geographical Review* 31(3): 353–364.
Sauer, C.O. 1956. The education of a geographer. *Annals of the Association of American Geographers* 46(3): 287–299.

Stanley, P. 2020. Unlikely hikers? Activism, Instagram, and the queer mobilities of fat hikers, women hiking alone, and hikers of colour. *Mobilities* 15(2): 241–256.
Stewart, K. 2013. Regionality. *Geographical Review* 103(2): 275–284.
Tindley, A., Gibbard, M., and Diamond, A. 2019. Archived in the landscape? Community, family and partnership: promoting heritage and community priorities through the Argyll estate papers. *Archives and Records* 40(1): 5–20.
Tolia-Kelly, D.P. 2016. *Landscape, Race and Memory: Material Ecologies of Citizenship.* Abingdon: Routledge.
Valentine, G. 2013. Tell me about …: Using interviews as a research methodology. In: Flowerdew, R. and Martin, D. Eds. *Methods in Human Geography: A guide for students doing a research project.* 2nd Ed. London: Routledge, pp. 110–127.
Vannini, P. and Taggart, J., 2014. Making Sense of Domestic Warmth: Affect, Involvement, and Thermoception in Off-Grid Homes. Body and Society, 20: 61–84.
Vannini, P. 2015. *Non-Representational Methodologies: Re-Envisioning Research.* London: Routledge.
Vannini, P. and Taggart, J. 2014. Making sense of domestic warmth: Affect, involvement, and thermoception in off-grid homes. *Body & Society* 20(1): 61–84.
Wang, C. and Kao, Y. 2017. Re-assembling the memorial landscape: The politics of walking tours in Taipei. *International Journal of Heritage Studies* 23(10): 1002–1016.
Wheeler, R. 2014. Mining memories in a rural community: Landscape, temporality and place identity. *Journal of Rural Studies* 36: 22–23.
Wylie, J. and Webster, C. 2019. Eye-opener: Drawing landscape near and far. *Transactions of the Institute of British Geographers* 44(1): 32–47.

Index